SOUTH POLE ODYSSEY

SOUTH POLE ODYSSEY

SELECTIONS FROM
THE ANTARCTIC DIARIES OF
EDWARD WILSON

Edited by
Harry King
Scott Polar Research Institute

BLANDFORD PRESS
Poole   Dorset

*Half of the royalties received by the Scott
Polar Research Institute from the sale of
this book will be used to provide Wilson
Memorial Grants to help young men and women
to undertake field-work in Polar regions. The
remainder will be used to assist the work of
the Institute.*

First published in the U.K. 1982 by Blandford Press,
Link House, West Street, Poole, Dorset, BH 15 1LL

Copyright © 1982 Scott Polar Research Institute, Cambridge,
England
Distributed in the United States by Sterling Publishing Co., Inc.,
2 Park Avenue, New York, N.Y. 10016.

British Library Cataloguing in Publication Data

Wilson, Edward
  South Pole odyssey.
  1. Wilson, Edward   2. Explorers — England — Biography
  3. Antarctic regions
  I. Title   II. King, Harold G.R.
  919.8'904   G875.W/

ISBN 0 7137 1206 6

Printed and Bound in Hong Kong by
Permanent Typesetting & Printing Co. Ltd

CONTENTS

Discovery with parhelia

INTRODUCTION

SOME SEVENTY YEARS have passed since the frozen bodies of Captain Robert Falcon Scott, Dr Edward Adrian Wilson and Lieutenant Henry Robertson Bowers were found in their snowed-up tent on the Great Ice Barrier of Antarctica, a grim conclusion to an enterprise that started so full of high promise. During these passing years the name of Scott has become a household word, his exploits symbolizing mankind at its finest when in adversity. Scott's published journals are still best-sellers. By contrast Edward Wilson is a shadowy figure; his published works consist largely of specialized scientific papers. A naturalist and doctor, Wilson accompanied Scott to the Antarctic on the *Discovery* expedition of 1901-1904. He returned with Scott a second time on the *Terra Nova* in 1910-1912, this time as his chief of scientific staff and right-hand man.

How greatly Scott depended on Wilson on the fatal expedition to the South Pole is evident from his journals and correspondence, which abound in praise of Wilson's personal qualities: 'Words must always fail me when I talk of Bill Wilson. I believe he really is the finest character I ever met — the closer one gets to him the more there is to admire.' Yet had it not been for the enthusiasm of a sympathetic and devoted biographer, the late George Seaver, it is probable that Wilson's name would have faded almost completely from the public view. Between 1934 and 1948 Seaver published three separate volumes covering different facets of Wilson's life, all based on material held by the family and Wilson's widow, Oriana, whose full confidence and cooperation in the use of diaries and correspondence he enjoyed. Seaver also obtained help and encouragement from a number of Wilson's former comrades, including Apsley Cherry-Garrard who was especially close to Wilson, and whose own book *The worst journey in the world* remains a classic of polar adventure to this day.

On the death of Oriana Wilson in 1945 the Scott Polar Research Institute in Cambridge inherited not only a fine collection of Edward Wilson's Antarctic drawings and paintings, including some of his original sketch books, but also some personal correspondence and his *Discovery* expedition journal bound in three handsome leather-covered volumes. Subsequent investigations revealed that much of

the material used by Seaver in his biographical research had disappeared, whether destroyed by Mrs Wilson or perhaps given away,
we do not know. However, it soon became apparent that in the
Discovery diary and in the wealth of accompanying Antarctic
pictures resided a potential publication which would be not only of considerable autobiographical value, but which might also serve to throw
further light on Scott's first Antarctic expedition. Sympathetically
reproduced, the sketches might bring before the public something of
the skill and the range of Wilson's exceptional pencil and brushwork.
Thanks to the interest of Blandford Press and to the enthusiasm and
industry of the Polar Institute's then archivist, Ann Savours, who in
turn received every support and encouragement from the Wilson
family, the *Diary of the Discovery Expedition to the Antarctic
Regions 1901-1904* was published in 1966, unabridged and adorned
with forty-seven reproductions of Wilson's watercolours. A year later
Blandford Press published a further volume devoted entirely to
Wilson's bird studies. The compiler was Brian Roberts, a world
authority in the field, and the book was published under the title
Edward Wilson's birds of the Antarctic. Designed to draw attention
to Wilson's genius for detailed recording in illustration, this
beautifully produced work assembled over three hundred of the
artist's drawings of penguins, petrels, skuas, albatrosses and other
birds of the Southern Hemisphere. Finally, in 1972, Blandford Press
arranged to publish Wilson's diary kept during Captain Scott's last
expedition, under the title *Diary of the Terra Nova Expedition to the
Antarctic 1910-1912*. This included the manuscript journal, held by
the Scott Polar Research Institute, covering the journey aboard the
Terra Nova to Antarctica and the period ashore up to the departure
for the South Pole, together with the diary kept during the South
Pole sledge journey, which is the property of the British Library. On
this occasion the present editor undertook to prepare the work for
press. Once again there was an opportunity to reproduce a further
selection of Wilson's Antarctic sketches, including those made at the
South Pole.

The idea of a 'cadet' or abridged edition of the *Discovery* and
Terra Nova diaries originated with Blandford Press, who felt that a
single volume highlighting the more adventurous episodes recorded
by Wilson would interest a wider public. The task of condensing two
sizeable volumes into one small one proved impractical. A better
idea, or so it seemed to this editor, would be to reproduce those
sections of the two diaries covering three major Antarctic journeys.
The first of these took place during the Antarctic summer of
1902-1903 and was the earliest attempt to sledge over the Ross Ice

Shelf, or the Barrier as it was called in those days, towards mainland Antarctica and the South Pole itself. On this occasion Scott, who was accompanied by Wilson and Lieutenant Ernest Shackleton, was away from the *Discovery* for ninety-three days. They navigated their sledge over the Barrier for a total distance of 960 statute miles, achieving a record southern latitude of 82°17' S and adding many newly discovered coastal mountain ranges to the map. All three men returned to the ship suffering from scurvy, Shackleton being in a desperate condition. For a number of reasons this first attempt to approach the South Pole fell far short of its objectives. All three participants were the merest beginners in the art of polar exploration, knowing little about dog-handling, surface conditions or diet. That they survived at all seems miraculous today. On the return to the *Discovery*, Wilson's close friend Shackleton was invalided home a very sick man; but such was his resilience that a few years later he was to launch his own attack on the Pole and come within an ace of success.

In 1911 Wilson found himself back in Antarctica, this time as Scott's second-in-command. Though the achievement of the South Pole was to be the climax of the expedition, Wilson was also determined that the interests of science should be served. Since the *Discovery* days it had been his ambition to make a trip in midwinter to the Emperor Penguin rookeries at Cape Crozier on Ross Island in order to collect newly-hatched eggs for research on the embryos. His account of that extraordinary excursion occupies the second extract from the diaries. This journey, on which Wilson was accompanied by the young assistant zoologist Apsley Cherry-Garrard and Lieutenant Henry (Birdie) Bowers, was probably one of the most hazardous field parties ever undertaken. During the five weeks that the three men were in the field — it was the middle of the Antarctic winter — they suffered from temperatures as low as -77°F. (-60°C.) and for much of the time were in almost total darkness. At Cape Crozier itself they almost lost their tent in a blizzard and survived, by the skin of their teeth, some hair-raising and roof-raising experiences in a temporary stone shelter. After an agonizing return trip they eventually reached Cape Evans more dead than alive, but with the object of the exercise intact — three Emperor Penguin eggs. As Captain Scott commented in his own journal: '. . . the result of this effort is the appeal it makes to our imagination, as one of the most gallant stories in polar history. . . It makes a tale for our generation which I hope may not be lost in the telling.' The unadorned, straightforward style of the narrative, completely lacking in any adverse comment on his companions and making light of all physical discomforts, is typical of

Wilson. He himself described the adventure as 'the weirdest bird-nesting expedition that has ever been made'. The low key of its style notwithstanding, the story makes gripping reading and makes a tale for our generation too.

In conclusion we reproduce in its entirety Wilson's field diary account of the dramatic and most memorable South Pole sledge journey from 1 November 1911, when he records the departure from Cape Evans of ten ponies and ten men, to 27 February 1912, the date of the last abbreviated entry written from the Barrier, four weeks before the author's death alongside Scott and Bowers. This South Pole diary is very different in style from the journals written up previously in the comparative comfort of winterquarters. The latter were fair copies, based on field notebooks and intended as a retrospective and rounded record of events largely for the information of friends and family. The Pole diary is itself a field notebook and is consequently more immediate and abbreviated in style. The original is now deposited in the British Library but the two sketch books which accompanied it are owned by the Scott Polar Research Institute. Found with the diary near the dead men by a search party on 12 November 1912, these sketchbooks are remarkable for the series of panoramic sketches of scenery which Wilson made as the Pole party laboured up the Beardmore Glacier.

With the exception of certain marginal annotations and sketches each of the episodes reproduced in this book is complete and unabridged from the Blandford Press editions of the *Discovery* and *Terra Nova* diaries. For the benefit of readers not familiar with these diaries or with the background to the two Scott expeditions, some explanatory notes have been included, together with a 'life' of Wilson and short biographies of all persons mentioned in the extracts. Each of the three episodes has been prefaced by a short foreword intended to explain the immediate circumstances leading up to Wilson's journal entries. With the exception of the South Pole diary, editorial policy has been to introduce paragraphs, to correct obvious mistakes — while retaining Wilson's idiosyncrasies — to restrict the use of capital letters and to explain abbreviations. The Pole diary, with its less considered logging of events, called for a slightly different approach. Too strict an adherence to the editorial rules might easily destroy the flavour of immediacy. It has been transcribed, therefore, with a view to reproducing the original as closely as possible.

Harry King
Cambridge, 1981

"DISCOVERY"

THE

SOUTH POLAR TIMES.

APRIL · 1902

Title page of first issue of *The South Polar Times*;
illustrated by Edward Wilson

Map of the Southern Hemisphere showing the main ports of call of
Discovery and *Terra Nova*

1

EDWARD ADRIAN WILSON

ORN AT 6 Montpelier Terrace, Cheltenham on 23 July 1872,
Edward Adrian Wilson was the fifth child and second son of
Edward Thomas Wilson, a highly respected medical practi-
tioner whose wife Mary, née Whishaw, came of a long line of Quaker
ancestors. The distinguishing characteristics of this sect — modesty,
industry and simple faith — seem to have been inherited by 'Ted', as
the family always knew him (in the Antarctic he came to be known as
'Bill'). Indeed the theme of Dr Wilson's unpublished life of his son,
written some years after the latter's tragic death, is the 'harmonious
evolution' of these traits of Ted's character from childhood to
manhood. At the gentle age of 18 months, Mrs Wilson was des-
cribing him as 'very loving and sweet, as a rule, but subject to violent
outbursts of temper and he will not brook contradiction.' Indeed the
boy's hypersensitivity led him to be easily moved to tears, and even in
adult life he was to find social engagements of any kind an agony.
These early emotional eruptions were certainly at the root of what
was later to be sublimated into a yearning to be of use to his fellow
creatures, at first in the practice of medicine and later in the self-
sacrificing devotion to duty and friends which so impressed Captain
Scott in the Antarctic.

Evidence was also forthcoming in Wilson's early childhood of the
artistic skills and the interest in natural history which, combined,
were to produce a scientific artist of no mean ability. At the close of
his third year, Ted's mother observed that he was 'always drawing'
and having artistic talents herself she began to give him lessons.
Alongside his interest in art Wilson began to take a keen interest in
nature. In 1874 his mother rented and began to farm the Crippetts,
an estate high up on a spur of the Cotswold Hills overlooking the Vale
of Gloucester. For the young boy, ever interested in beetles,
butterflies and birdsnesting, it was a perfect heaven. Over the years
he came to know every stick and stone of it. Here it was that he
learned to develop the gifts of accurate and patient scientific
observation of animal life which were to mark him out in later years
as a field-naturalist of great talent. Wilson's special love was for
birds. It was said of him that not only could he recognize any bird by

its song, but that he could predict its particular occupation at that particular moment. Throughout his life, Crippetts' woods offered peace and solace from the stresses and strains of a working life, and in the Antarctic he would find relief from the pain and tedium of sledging over a frozen wilderness by returning, in his imagination, to that leafy Gloucestershire paradise.

At the age of fourteen Wilson entered Cheltenham College as a day-boy, and it was here that he began to keep a journal, inspired perhaps by reading Darwin's *Voyage of the Beagle*. It was to begin as a much abbreviated affair; the events of a week would occupy a single page, each day being concerned with such events as 'caught a young hedgehog toddling up the road 7.45 p.m.'

In 1891 Wilson went up to Caius College, Cambridge, where he read for the Natural Science Tripos with a medical career in view. His father described him at this time as 'reserved, with few friends and many acquaintances, aesthetic and very religious.' His boyhood habits of birdsnesting, photography, collecting and sketching persisted and his college rooms were described by a contemporary as 'more like a museum than a dwelling place'. Wilson also developed a taste for art, as interpreted by John Ruskin, and for the works of Tennyson, whose *In memoriam* he was to read and re-read in the Antarctic — 'a perfect piece of faith and hope and religion' was how he described it in his Pole diary. As a sport he took up rowing, an activity which, as his friend Cherry-Garrard was later to point out, requires much the same qualities as sledging.

Finally, in 1894, Wilson took his B.A. degree and ended what he called the 'buttefly' existence of the academic life by leaving Cambridge to enter St George's Hospital, London, as houseman. His medical career had begun in earnest. To begin with he made a special virtue out of poverty, living frugally on potatoes and coffee while smoking 'to excess'. For a while he lived at the Caius College Mission in Battersea, doing good works in his spare time, especially among the poor children of the neighbourhood. On top of this and a strenuous medical course, Wilson found time to read and to absorb the works of his mentor Ruskin, to sketch animals at the London Zoo and to visit the National Gallery where he was 'smitten to distraction' by Turner's drawings. Letters home from about this time show evidence of growing religious convictions; indeed, such were their strength that, at one point, he was within an ace of volunteering for missionary work in Africa.

Then, in 1898, came the crisis which was to change the whole course of Wilson's life. Under the strain of overwork — he was studying for his Cambridge M.B. on top of his many other duties —

14

his health broke down. Tuberculosis was diagnosed and he was packed off, at first to Norway, and afterwards to Switzerland, to recuperate. In the pure crystal air of these retreats he was able to rest and indulge his talents as an observer and artist; the snow, the mountains and the wildlife especially appealed to him. It was here, too, that he began to lay down the foundations of a personal mysticism, from which later stemmed the selfless reliability which enabled him to sustain not only himself but also his friends in times of crisis. Just how private was his personal faith in Providence, was commented upon by Cherry-Garrard years later on board the *Terra Nova*: 'You must not think of Bill as a religious man. It has come almost as a shock to some of us to learn now for the first time that he held a Service to himself up in the crow's nest every week.'

In 1899 Wilson returned home, his health improved but still precarious. It turned out to be a memorable year, for in October he became engaged to Oriana Souper (Ory) daughter of a Huntingdonshire parson whom he had first met during his Caius Mission days. Then, in December, he travelled up to Cambridge to sit his medical examinations and, despite the gloomiest forebodings, learned that he had passed. The next year, 1900, turned out to be even more momentous. Wilson began to devote an increasing amount of time to bird drawing, which he now came to regard as a vocation, until such time as his health would allow him to take up a hospital appointment. It was also the year in which a young Royal Naval officer, Robert Falcon Scott, recently appointed to command the National Antarctic Expedition by its joint sponsors, the Royal Society and the Royal Geographical Society, was looking for a candidate to fill the post of junior surgeon and zoologist. A member of the Appointments Committee was Philip Sclater, President of the Zoological Society of London, who knew Wilson's skills as a scientific draughtsman and was anxious for him to apply for the post. Wilson's uncle, Sir Charles Wilson, was also keen for him to go and brought his talents to the attention of the promoter and 'father' of the expedition, Sir Clements Markham, the formidable President of the Royal Geographical Society. Sir Clements was quick to recognize Wilson's potential value to the expedition and advised him to seek an interview with Scott. In November, suffering from a painful outbreak of blood poisoning contracted while carrying out a post-mortem examination, Wilson attended for interview and was accepted subject to a satisfactory medical report. By the summer of 1901 he had been pronounced fit and joined the expedition as surgeon, artist and zoologist. On 6 August 1901, only three weeks after his marriage to Oriana Souper, Wilson sailed for Antarctica on board *Discovery*.

Prophetically he had written home: 'I am going. They accept me in spite of everything if I go at my own risks. I don't care in the least if I live or die — all is right and I am going; it will be the making of me.'

The full story of the *Discovery* expedition can be read in Scott's *Voyage of the Discovery* and in Wilson's own *Diary of the Discovery Expedition*. It was the first attempt by a wintering party to carry out an extensive scientific exploration on the Antarctic mainland. A single season had originally been envisaged, but, because of bad ice conditions, it was found impossible to free *Discovery*, used as winterquarters, from her moorings and the ship's company was compelled to spend a second winter down south, eventually returning to England via New Zealand in 1904. Among the numerous field trips planned by Scott to explore the coast line and hinterland, by far the most ambitious was the Southern Journey, the first serious attempt to find a route to the South Pole. In chapter 2 we reproduce Wilson's own account of this remarkable feat. In it, he describes how with Ernest Shackleton, a young merchant navy officer, Scott and he achieved what in those days was a record southern latitude of 82°17', and tells how the three of them finally struggled back to *Discovery* after narrowly escaping ultimate disaster due to scurvy and near starvation.

On his eventual return to England in September 1904 Wilson was again obliged to face up to the problem of his future career. Lecturing to schools and scientific societies, the writing up of his scientific notes for publication, and the exhibition of his pictures occupied every hour of his free time for several months. Then, in 1905, the post of field-observer to the Board of Agriculture's Commission on the Investigation of Grouse Disease was advertized and Wilson was appointed. The work was to occupy him on and off for nearly five years and involved him in research on the cause of an unknown disease that was killing Scottish grouse in large numbers. The grouse disease investigations were precisely suited to Wilson's training as a doctor and field ornithologist in Antarctica, and had the advantage of offering him six months' employment on the Scottish moors while leaving him the winter months to sketch and write. Nevertheless, it was demanding work and occupied him so fully that he was still busy writing up the results halfway to Antarctica in 1910. Wilson personally dissected some 2,000 grouse before locating the root cause of the disease, a minute threadworm. Besides the task of preparing the colour plates for this report, Wilson was also commissioned to illustrate a book of British mammals and a book of British birds; and at the back of his mind were plans for a book of his own drawings and watercolours. It was in the nature of the man to take on far more work than he could comfortably manage. There were, however, some treasured moments of relaxation shared with his wife Oriana, including visits to Cortachy, Kirriemuir, in Scotland, where their friend Reginald Smith, the publisher, had a small hunting lodge. Here, in 1907, they were joined by Captain Scott who even then was making plans for a return visit to the Antarctic.

British interest in Antarctic exploration, which had been in the doldrums since the return of the *Discovery* expedition, was reawakened by an announcement in the press in February 1907 that Lieutenant Shackleton was to attempt the South Pole with the financial backing of the Scottish industrialist Sir William Beardmore. The obvious candidate for the post of second-in-command was Shackleton's old friend and sledge mate Bill Wilson. Shackleton used all his powers of persuasion to entice Wilson into his service: 'I want the job done and you are the best man in the world for it,' he wrote, 'and if I am not fit enough to do the southern journey there could be no one better than you...Come, Billy. Don't say no till we have had a talk. Don't say no at all...' This invitation Wilson rejected firmly and peremptorily. He had, in fact, already intimated to Scott that he would stand by to return with Scott to the Antarctic when the need arose.

Scott's own reaction to Shackleton's plans was one of irritation. That Shackleton should envisage the use of Scott's former winter-quarters at McMurdo Sound seemed to Scott tantamount to a breach of professional etiquette. It appeared as though a first class row might well have broken out had not Wilson himself intervened to pour oil on the troubled waters. Shackleton was persuaded not to operate from McMurdo Sound unless forced to do so by circumstances beyond his control. In the event Shackleton found himself unable to set up a base on the Barrier, at the Bay of Whales, as planned and was compelled to break his word and establish winter-quarters at Cape Royds on Ross Island. The whole episode had the effect of moving Scott to advance his own plans. In September 1908 he wrote to Wilson making him a firm offer of the post of Head of the Scientific Staff. In a letter to his mother Wilson confided his reasons for accepting: 'For my own part I have long been convinced that the first principle of right living is to put one's life into the hands of God and then do the work He gives one to do... There has been no choice, and therefore no difficulty to my mind in deciding...'

In June 1909 Shackleton returned to a hero's welcome and a knighthood. He had discovered and successfully climbed the Beardmore Glacier, the route to the polar plateau and the South Pole itself, the ultimate goal, which in the end eluded him, though by only 97 geographical miles. At a political dinner held in Shackleton's honour, Scott declared that 'the Pole must be discovered by an Englishman,' and a *Times* leader subsequently declared that, 'it would be deeply regrettable if for want either of men or of money, the brilliant recent record of Antarctic exploration were at this point to be checked, with the inevitable probability in such a case that the Pole would first be reached by an explorer of another nation.' Rumours were indeed circulating that French, German and Japanese parties were preparing to take to the field, and that the American, Robert E. Peary, conqueror of the North Pole, was scheming to plant the Stars and Stripes at the South Pole. Scott hesitated no longer. On 13 September 1909 he opened an office for the British Antarctic Expedition at 36-39 Victoria Street, London, and a few days later asked Wilson to organize and lead the scientific staff, an offer which the latter was quick to accept.

During the months that followed, Wilson worked round the clock selecting his various specialists, for, as he wrote to his father: 'We want the scientific work to make the bagging of the Pole merely an item in the results.' During this time Wilson was consulting and being consulted on every aspect of the expedition's organization, including the calculation of food values and quantities, and on the selection of scientific instruments. On top of all this Wilson was still engaged on the Grouse Disease Report and the illustrations for the books on British birds and mammals. His strength lay in a superb self control rooted in a deep religious faith. 'Free will,' he wrote, 'in the eyes of God means the willingness to do one's best at whatsoever comes in our way and however difficult it may seem... He does not like us to look back after putting our hand to the plough. He often takes the plough away as soon as He knows we mean to carry through. Nothing can disappoint one if one sees things in this way, because everything has been done in the eyes of God if the intention is there to do it... In this way one lives a long life in a few years.' The practical application of these beliefs to the problems of Antarctic exploration is in evidence throughout Wilson's journals. His account of the successful accomplishment of the midwinter journey in July 1911 is a perfect instance of the mastery of mind over matter.

By the late spring of 1910 all was in readiness for the departure of the expedition in the *Terra Nova*, a former Labrador sealer purchased by Scott to take the place of the *Discovery*, by then the property of the Hudson's Bay Company. Wilson somehow managed to squeeze in a few weeks aboard a whaler off the Shetland Islands learning at first hand and in considerable discomfort techniques which he hoped to apply in Antarctic waters for scientific research. Finally, on 15 June, the *Terra Nova*, escorted by an armada of small boats, sailed from Cardiff docks. 'Everyone', wrote Wilson's father, 'was full of hope, tho' sad at heart, but not without foreboding; hope was uppermost.' Life on board ship during the long voyage south was, for Wilson, an opportunity to experience the routine that so suited him in the *Discovery* — painting endless sunrises and sunsets, penning letters home, writing up his diary and scientific reports and taking part in the boisterous camaraderie of the wardroom; for 'Uncle Bill', as he was familiarly known, with all his religious convictions and high moral standards, was no prig or kill-joy. Less agreeable were the stints of duty trimming coal and stoking the ship's boilers, an especially disagreeable task in the tropics.

Eventually Cape Town was reached and Wilson was reunited with his beloved Oriana, who had gone ahead of him by fast passenger liner. After a brief holiday exploring the surrounding veldt and countryside, Wilson, albeit with reluctance, agreed, at Captain Scott's urgent request, to proceed by mail boat to Australia, there to recruit scientific staff and raise funds from the Australian Government; for the expedition was seriously short of funds. Money raising was not a task congenial to Wilson's retiring nature, but he saw the business through and enjoyed to the full the extra time it afforded him to be with his wife.

In October Wilson once again rejoined the *Terra Nova* at Melbourne for the voyage to New Zealand. Here the ship was overhauled and the members of the expedition were entertained until the time came for the final departure from Dunedin for Antarctica on 29 November. Oriana Wilson spent the last hours with her husband on board the *Terra Nova*, until, as he noted in his diary, 'she had to go off on to a tug — and there on the bridge I saw her disappear out of sight waving happily, a goodbye that will be with me till the day I see her again in this world or the next — I think it will be in this world and some time in 1912.'

From then onwards the course of Scott's last expedition was beset by a series of drawbacks and adversities. Having all but foundered in prodigiously high seas south of New Zealand, the *Terra Nova* finally entered the pack ice, which bars the approaches to the Antarctic continent, only to be delayed for over three weeks by unseasonably heavy concentrations of ice. She emerged at last into the open Ross Sea, but on reaching Cape Crozier on Ross Island, the expedition's intended base, it was found impossible to land owing to the heavy swell; Scott had to make do with the much less advantageous site at Cape Evans on the opposite side of the island. Here stores had to be off-loaded on to the frozen sea ice of McMurdo Sound. In the course of this operation, one of three experimental motor sledges, on which Scott was pinning some hopes as auxiliaries for a successful dash for the Pole, dropped through thin ice on to the sea bed. The remainder turned out to be too mechanically unsound to be of service on the depot-laying journey. This operation, carried out in some haste because of the impending onset of winter, was similarly beset with mishaps. Its purpose was to support the forthcoming Pole expedition by depositing caches of stores, fuel and pony fodder as far to the south as possible for the use of the outgoing and returning Pole parties. The depots were marked by cairns of snow topped by flags. Scott planned for the bulk of the stores to be hauled on sledges by ponies, supplemented by dog sledges. Unfortunately the Siberian ponies, of which Scott had high expectations for hauling heavy loads, were in poor shape and did not easily adapt to Antarctic conditions. Out of eight employed on this exercise only one survived, leaving a total of ten for the final venture. Scott's hopes of a southerly depot at latitude 80° were disappointed, instead a ton of stores (One Ton Depot) were deposited at latitude 79°28½ ' S. And then, to cap all these setbacks, came the stunning news that a rival Norwegian expedition, led by the formidable Roald Amundsen, successful navigator of the Northwest Passage, had set up a base at Framheim on the Barrier to the east of McMurdo Sound, sixty miles closer to the South Pole.

Wilson's reactions to these vicissitudes, as revealed in the pages of his diary, were characterized by quiet resignation to God's will. Of the pony disaster he commented: 'I firmly believe that the whole train of what looked so like a series of petty mistakes and accidents was a beautiful prearranged plan in which each one of us took exactly the moves and no others that an Almighty hand intended each of us to take — and no others... The whole thing was just a beautiful piece of education on a very impressive scale.' Wilson was critical of Amundsen's prospects of achieving the Pole, but was not unduly worried should the Norwegians, after all, gain the advantage: 'He [Amundsen] may be fortunate and his dogs may be a success in which case he will probably reach the Pole this year, earlier than we can, for not only will he travel much faster with dogs and expert ski runners than we shall with ponies, but he will be able also to start earlier than we can as we don't want to expose the ponies to any October temperatures.'

Wilson's natural tendency to avoid dramatization of events, and especially to eschew self-advertizement, is especially evident in his account of the mid-winter journey to Cape Crozier in July 1911. A major crisis in which a gale-force wind carried off the roof of a makeshift shelter exposing Wilson and his two companions to the howling elements is glossed over as 'quite the funniest birthday I have ever spent', and an accident involving a spurt of boiling oil in one eye is dismissed in a single unemotional sentence. Wilson's Pole diary is of necessity terse and matter-of-fact, and his reactions to such moments of high drama as the discovery of Amundsen's priority to the Pole are utterly unemotional: '. . . he can claim prior right to the Pole itself. He has beaten us in so far as he has made a race of it.' Of Petty Officer Edgar Evans' collapse and death on the return down the Beardmore Glacier Wilson remarks simply: 'Evans' collapse has much to do with the fact that he has never been sick in his life. . . '

Wilson's Pole diary ceases abruptly on Tuesday 27 February 1912 when with Scott, Bowers and Oates he was about to commence the final march across the Barrier towards One Ton Depot and, hopefully, Cape Evans. It was a significant event. Since his Cheltenham College days Wilson had kept up his habit of diary writing even in the most adverse Antarctic conditions. Now he was compelled to lay it aside. Clearly rapidly deteriorating circumstances had forced him to abandon what normally he would have regarded as a prime duty. From then onwards one can only catch glimpses of the man through the memorable last pages of Scott's own journal: 'Wilson, the best fellow that even stepped, has sacrificed himself again and again to the sick men of the party. . . We none of us expected these terribly low temperatures and of the rest of us Wilson is feeling them most; mainly, I fear, from his self-sacrificing devotion to doctoring Oates' feet.' On 16 or 17 March, Oates, that 'very gallant gentleman', went out into the blizzard never to be seen again. On 21 March, Wilson, Scott and Bowers, suffering fearfully from frostbite, set up their tent for the last time. Eleven miles away, at One Ton Depot, lay the fuel and food that might have been their salvation. But the continuing and relentless blizzard prevented any attempt to reach it. On or about 29 March Scott concluded his own diary with a final *cri de coeur*: 'For God's sake look after our people.'

As he lay in the tent awaiting the inevitable, Wilson took up the pocket sketch book in which he kept his diary to write last letters home, clinging to the last shred of hope that some search party would come upon the letters and take them to their destination. One of these, to his father and mother in Cheltenham, epitomises Wilson's fortitude in adversity and his abiding faith in the life-to-come: 'The end has come and with it an earnest looking forward to the day when we shall all meet together in the hereafter. Death has not terrors for me. I am only sorry for my beloved Ory and for all of you dear people, but it is God's will and all is for the best. Our record is clear and we have struggled against very heavy odds to the bitter end...'

It was little short of a miracle that the bodies of the three men came to be found by a search party led by Surgeon Edward Atkinson, commanding the wintering party at Cape Evans, some eight months later. The date was 11 November 1912. Atkinson, the first to enter the tent, noted that 'Bill had died very quietly with his hands folded over his chest'. The reaction of Tryggve Gran, the young Norwegian ski expert, who followed Atkinson inside, was rather more dramatic. What he saw was to remain with him for the rest of a long life: 'Dr Wilson', he tells us, 'was sitting in a half reclining position with his back against the inside of the tent facing as we entered. On his features were traces of a sweet smile and he looked exactly as if he were about to wake from a sound sleep. I had often seen the same look on his face in the morning as he wakened, as he was of the most cheerful disposition. The look struck me to the heart and we all stood silent in the presence of this death.'

Note: Temperatures are given in Fahrenheit. The freezing point of water is 32°F. Distances are given in either geographical or statute miles.

1 statute mile = 1760 yards

1 geographical mile = one minute of a great circle of the earth which varies owing to the geoid form of the earth and has been standardized at 6080 feet, its value at latitude 48°. 1 geographical mile = 1.15 statute miles.

5 statute miles = 8 kilometres

Edward Adrian Wilson (1872-1912)

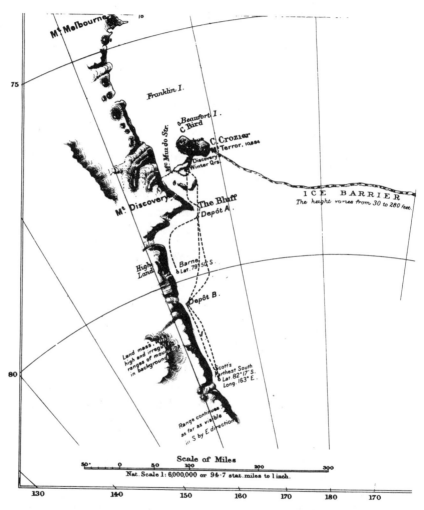

Detail from a contemporary sketch map showing the route of the
southern party, 1902-1903.

2

THE SOUTHERN JOURNEY, SUMMER 1902-1903

PROLOGUE

IT WAS ON 12 June 1902 in the depth of *Discovery's* first Antarctic winter that Captain Scott called Bill Wilson into his cabin and revealed his plans for the forthcoming spring sledging parties. The most ambitious of these was to be a long journey southward towards the Pole which Scott himself would lead. It was to consist of two or three men and all the available dogs. To Wilson's surprise, especially in view of his own medical history, Scott asked if he would accompany him. Wilson agreed, but argued strongly for a third man so that if one of the party fell ill there would be a better chance of all getting back. Scott readily concurred and to please Wilson suggested as the third man Lieutenant Ernest Shackleton ('Shackles'), a young merchant navy officer and Wilson's best friend on the expedition. Wilson knowing that it was Shackleton's burning ambition to be included on the southern journey felt bound to agree. But by the following September he was beginning to have private doubts about the whole enterprise. There was the matter of his own scientific research, in particular a study he was planning of the breeding habits of the Emperor Penguin, a vast rookery of which had been discovered at Cape Crozier by Lieutenant Reginald Skelton. This work would either have to be abandoned or handed over to Wilson's immediate superior, Reginald Koettlitz, the senior medical officer. Worse still, Wilson was deeply concerned for Shackleton; '. . . for some reason,' Wilson wrote, 'I don't think he is fitted for the job . . . he is so keen to go, however, that he will carry it through.'

On 2 November the Southern Party, with nineteen dogs, five sledges and a supply of food for nine weeks, set off cheered on their way by the whole ship's company. Three days previously a depot party of twelve men, under Lieutenant Michael Barne, had gone ahead to manhaul extra supplies that would see the Southern Party through a further four weeks' sledging if need be. The route south lay over the undulating surface of what, in those days, was called the

Great Ice Barrier, or simply the Barrier (today known as the Ross Ice Shelf). About the size of France, this seaward extension of the vast ice sheet which mantles the Antarctic continent, and is partly afloat and partly grounded, provides a route from the Ross sea coast to the distant mountains whose glaciers give access to the high polar plateau surrounding the South Pole. With the twin volcanoes of Erebus and Terror on Ross Island dominating the scene, and to the west the towering peaks of the Royal Society Range on mainland Antarctica, there would at first be no lack of magnificent scenery for Wilson to sketch. Later, after passing ice-embedded White Island and the promontory known as the Bluff, the three men would be making for the hitherto unexplored coastal ranges and a possible route to the Pole. Theirs was not, however, the first attempt to sledge on the Barrier. In February 1900 a British expedition led by a Norwegian called Carsten Borchgrevink, in command of the ship *Southern Cross*, had succeeded in making a landing there, and had then succeeded in making a short trip over the ice to latitude 78°50′S, a record surpassed and duly celebrated on 12 November 1902 when Scott, Wilson and Shackleton bade farewell to Barne's support party and headed towards the unknown.

DIARY

Sat 25 Oct   I hear we are to start on our southern journey the middle of next week. The Skipper told me this when I asked him if I could have three or four days out northward in our strait to hunt for Emperors here, now that we know how they breed and the likely place to look for them. I think they may quite possibly be breeding on the shore ice around the glacier we visited. They must lay their eggs very early indeed and carry them till they hatch and then continue carrying the young. The eggshells they swallow. The Skipper says there is not time for me to go and hunt for them, but Koettlitz is going instead. I am afraid this long southern journey is taking me right away from my proper sphere of work to monotonous hard work on an icy desert for three months, where we shall see neither beast nor bird nor life of any sort nor land and nothing whatever to sketch. Only I think we must, and hope we shall, come to land when we have travelled south on the Barrier for a month or so. Anyhow it is *the* long journey and I cannot help being glad I was chosen for it. *If* we come across anything but Barrier, it will be exceedingly interesting.

Sun 26 Oct   A day of Antarctic summer, quite wonderful in its way, almost like Davos[1]. Temp. −15°F. but dead calm and a hot bright sun in a clear sky. Church as usual in the morning and then the Skipper explained the use of a theodolite to me on the floe. The men play football and amuse themselves on ski, coming frightful croppers down the steeper snow slopes sometimes, though many of them are regular experts.

After tea I went on ski round to Pram Point, where a large number of seals have been coming up lately, for a week now, and young ones are being born. We found two infants with their mothers, one had been born that morning, the other has been very freely photographed for the last five days. It was born on Wednesday last. Both had their eyes open. They bah'd like sheep, and the mothers were inclined to make a dash at one if one touched the infants. Their coats were rough and furry, but not woolly, and in colour and marking they were much like the adults only paler, no black, more rusty brown all over. There were quite a number of cows here that would shortly give birth, and many bulls that had been fighting and were bitten all over the neck, chest and belly. One of the cows, seen today at Castle Rock, was terribly gashed on the underparts, probably having escaped an *Orca* (Killer Whale) as open water is within a measurable distance and the whales are there, as we know from the fact that the Pilot[2] saw them when he reached the edge of the ice on his sledge journey. I got home in time for dinner at 6.

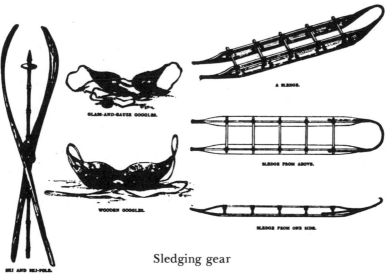

Sledging gear

SLEDGING PANNIKIN AND SPOON.

SLEDGING HARNESS.

FINNESKOES.

It is a treat having Royds back, as we now get the piano played. One gets a little wearied by the hammering of the pianola day after day. The midnight sun was visible today.

Mon 27 Oct   From breakfast until dinner was painting the young Emperors. After tea, which is now our midday meal at 1.30, I started skinning them with Cross and we finished very soon after 6, making two very good skins. One was a male, the other a female. Their length from the tip of the beak to the tip of the tail about 13 inches. After dinner writing and sewing.

Tues 28 Oct-Wed 29 Oct   Both these days spent entirely in preparing things for the sledge journey and getting everything ready for a start. It means a lot of sewing and a lot of altering. Buttons all sewn on with twine; ski, fur boots or finnesko, ski boots, burberries, headgear, hand gear, sleeping gear, all want an amazing amount of attention and thought. For on attention to details in them depends not only one's comfort but safety. Nearly the whole of Wednesday I spent clearing out my cabin and stowing away every available thing, all my papers and oddments, in the drawers, that the room may be scrubbed down from top to bottom. The ceiling and walls are thick with black from the lamp, as is everything in the cabin. The dirt is simply horrible, but one has got more or less used to it.
The weather has been warm, hovering a little above zero, and a great deal of snow has fallen, an indication of summer weather. The black bulb today showed no less than 126°F. in the sun.

Thurs 30 Oct Snow showers flying around and at times a good deal of wind and gleams of sunshine. Michael started off today in advance of us, with a party of twelve men. As a matter of fact nearly everyone on the ship went off with them and gave them a hand. There were flags galore — a Union Jack, Michael's banner — an Irish flag with a harp on a green ensign — and a composite flag made by the depot party for themselves with three mottos, 'Hope on, hope ever,' 'Now we shan't be long,' 'No dogs admitted.' The last referring to their party being man-haul, whereas ours, which starts now on Saturday is dog-haul. The Captain, Royds and I went with them as far as Cape Armitage thermometer, which is a mile and a half from the ship.

All the afternoon went in writing directions as to what was to be done in the way of seals and birds during my absence. We now have sunshine all day and night and it makes a big difference. The weather is unsettled but on the whole much warmer, this morning plus 13°F. and the snow falls soft instead of like sand grains. The evening went in writing a few letters home.

Fri 31 Oct All day spent in final arrangements and preparations for our start early tomorrow, labelling and burying young seals and seal skins in snow on shore with Cross, and making an inventory of my bird skins etc. There are now some fifty skins in all. This took most of the morning and the afternoon went in sewing etc.

The evening I spent writing to Ory, interrupted by callers who were dropping in all the evening to have a talk and wish us good luck and a long record journey south. Can anyone, I wonder, realize exactly what it is, leaving the ship and all one's companions, except two, for three months in this desolate region to walk down into the absolutely unknown south, where as far as one can see, nothing awaits one but an icy desert and one literally carries one's little all on a sledge! It's a funny game, because one has got so very attached to the ship as a home and the whole bay here and hut as a sort of estate. One's cabin moreover is full of one's home associations, a small sanctuary for happy recollections, lamp soot and general comfort.

Sat 1 Nov Blowing a regular blizzard, making it quite impossible to start. Hanging about the whole morning intending to start the moment it clears. Tea time arrives 1.30 and still we are here. So we have a treat given us by Dr. Koettlitz in the shape of a full plate of fresh cut mustard and cress each. After tea still impossible to start, though clearing and as soon as we were fully equipped for a start, up comes blizzard stronger than ever, so we undress again and hope on. Having it in our minds to start even tonight if it clears, we all lay down and slept the afternoon away, but at dinner it was blowing as hard as ever with a blinding drift of snow. So the usual evening cards are settled down to and I write my diary and another line to my Ory.

Sun 2 Nov Twenty third Sunday after Trinity. At last we started, though it was a cold and windy day. About 10 a.m. the Captain, Shackle and I were photographed with our nineteen dogs and five sledges and our flags. It was a great send off by the whole ship's company and we started at a great pace, accompanied by many of the others — Koettlitz, Skelton, Bernacchi, Hodgson, Royds and others. We had 1700 lbs weight at starting. We also carried 12 pair of ski and ski poles for Michael's party, who are ahead of us and who we think would manage better on ski than as they are. Very heavy squalls of drift and whirlwinds the whole morning, fewer during the afternoon.

At 2 p.m. we pitched tent at the old autumn depot and made some tea, having covered between 6 and 7 miles. We passed one used seal hole which must have been on the new ice, but there was no seal up and the line of demarcation between the old and new ice was not apparent today, where we crossed it, being shallow and drifted level at that spot I imagine.

The light was bad. There was low drift everywhere, overcast day but clearer to the south. At one time a whirlwind of drift swung right into our midst and for a few minutes we couldn't see one another at all. We caught up Michael's party at the northwest end of the White Island, they having just struck camp and started off again. We gave them the ski to try, but their efforts were absurd. They couldn't budge the weights they had on their sledges, as the surface was very irregular, covered with high smooth slippery sastrugi[3]. Nell, one of the bitches, slipped her harness and got away soon after we camped for the night and as she has pups on board we are afraid she will not come back to work with us. I am not sorry, though it reduces our limited supply of dogs. Everything else went well and there is now no wind and we are in our sleeping bags, 9 p.m. having done a long day's march.

Mon 3 Nov Called at 6 a.m. but before we were out of the tent a blizzard came on and lasted till 8 when we made breakfast and started off, able to see depot party as a black speck on ahead. Wind southwest and after a very tiring march against a head wind we caught them up at 2.30. Temperature in the tent rose during

cooking to plus 75°F.! Outside it dropped to zero, but rose after midnight to plus 10, remained plus 5 about most of the day.
Surface very irregular, hard and slippery then soft, straining one's legs horribly by constant slipping in finnesko[4]. Many sastrugi all denoting southwest prevailing winds. The dogs' pace is fast at present and one often has to run to keep up, sometimes on soft snow, sometimes on icy wind ridges. I feel badly strained in knees, hamstrings and heel tendons. When we caught up the depot party they camped with us for lunch. Michael lunched in our tent. At noon we lost Observation Hill and opened up the Bluff to the south of White Island. No sun all day till late in the evening. The depot party had caught Nell and returned her to our team. After tea and a wait until 5 p.m. we journeyed on and again caught up the depot party by 7 when we camped. Our sleeping bags are wonderfully comfortable. With no wind these temperatures are perfect for sledging. Turned in at 9 p.m.

Tues 4 Nov   The other party camped ¾ of a mile ahead of us over night, but were well away before we struck camp at about 10 a.m. We were not under way much before 12, but caught them up camped for lunch at 2 p.m. so we also had lunch. The going was very soft in places and very slippery in others. I was leading all the morning and felt the strain in the knees and hamstrings badly. After lunch the Captain and I went on ski and found them a great relief. Shackle led. Weather fine, a cold head wind but temperature above zero all day. Last night − 2.5°F. was the lowest. No drift today worth mentioning. We made 8 geographical miles and opened out Mount Discovery at 2 p.m. No ridges or bad pressure hummocks visible to us, but we were crossing long low waves which periodically hid the depot party from our sight. We crossed these waves about one to the mile. At lunch we change footgear, as we are wearing ski boots and one cannot sit for an hour in the tent in them with safety. We slip on our night finnesko (fur boots). The dogs pulled well today.
I am bothered with a very acute cold in the head, and aching pains, which one hardly expects on a sledge journey.

Wed 5 Nov   Camped with the depot party who are in splendid form and high spirits. Quiet night with no wind and bright sun. We turned out as soon as the other party were well away. They were again dipping into long valleys and so disappearing from sight. We started off on a splendid surface and in splendid weather. Fine halo round the sun and a nasty cloud forming over the Bluff, which as a rule means wind. This soon came on from southwest blowing up drift in our faces, which often rose sufficiently to obscure the party ahead of us. Soon we saw them camp and shortly we came up with them and camped also. We got into our bags and I read aloud a chapter out of the *Origin of Species* which we had brought for these occasions. We discussed it between whiles, and at 6 p.m., as the wind and drift continued, we made supper and settled in for the night. The weather looks very threatening, but there is less wind. The sledges are buried in snowdrift, dogs as well. Temperature plus 8.5°F. My cold much better. Aching pains all over as soon as one gets warm in the sleeping bag.

Thurs 6 Nov We had a disturbed night, though a long one. The dogs, having had little work, were restless and three times we were out to stop fights and tie up dogs that had got adrift. Sky overcast all night, cleared towards morning. Beautiful sunshine then and sky clear for all but a little cirrus. Started at 10 a.m. Soon caught up and passed the depot party and went on right ahead of them. Passed over some very high and well marked ice ridges of pressure and a great number of old snowed up bridged crevasses, from a foot or two to many yards across, all running northwest to southeast. Camped for lunch about 2 p.m. less than an hour. Sun very hot and all of us sweating freely. Socks drying well in the sun. Sledges overturned pretty often today going over high wind swept ridges. Very rough travelling. After lunch we came to an area of very high sastrugi all running southwest to northeast, showing the prevailing wind to be southwest round the Bluff. The view is very grand now: Erebus and Terror gradually getting smaller and smaller, White Island following suit, the Bluff standing out black and bold and rocky against Mount Discovery. We went on till nearly 7 p.m. and covered about 13 geographical miles in the day. Camped in glorious sunshine, perfect weather, no wind, temperature plus 8°F.
My cold almost gone, feeling fresh and fit at the end of the day's work. Shackle started a most persistent and annoying cough in the tent.

Fri 7 Nov Breeze sprang up from the southwest immediately we had turned in and a low bank of stratus came up, betokening a blizzard, which increased all night and was very thick and strong at 11.30 when we turned out and had our breakfast. As soon as this was over, we got into our bags again and Shackle read a chapter of Darwin. I did some sewing. We lay low in our bags talking, sewing, reading and sleeping off and on till 4 or 5 p.m. when the blizzard began to thin down. So we made ourselves some tea, dug out the dogs and sledges and started off about 9 p.m.

From 9 till 1 a.m. we marched over very smooth, slippery and high sastrugi with a gentle northerly breeze behind us. Temperature 3° or 4° below zero all the time. Ski boots never got soft, but we have the sun ahead of us in the south very bright and the weather looks more promising. Sledges capsized fairly often over the high wind ridges. Very bad going. The dogs pulled well, poor beasts. Camped at 1 a.m. and were in our bags after supper by 4 a.m. Fine and clear sun. Socks all hung out to dry. Feel very fit indeed myself, but Shackle is coughing a great deal. Young moon just visible. Every prospect of a fine day.

Pitching tent in a high wind

Sat 8 Nov   At 10 a.m. it was again blowing and drifting from the southwest and though the sun remained and the sky was clear, the wind kept up to force 3 or 4 (Beaufort's scale) and maintained a drift, which being dead against us prevented our moving, so we lay low in our bags all day. Breakfast at 2.30 p.m. Supper at 8 p.m. and at 10 p.m. we settled in for the night. Still blowing and drifting. Had to dig ourselves out of the tent door, sledges nearly out of sight. All the same we had a decent day. I read a chapter of Darwin. There is as usual an ugly cloud cap over the Bluff and a heavy bank of cloud coming up from the southwest.

Here ends our first week out. We have covered about 50 geographical miles since we started. I may say here that miles in future always mean geographical and not statute miles. One geographical mile here measuring 1025 fathoms or 2050 yards instead of 1760 yards, which is the statute mile. There are probably a good many things mentioned which want some explanation, but I think I had better make them as side notes when I have finished copying these notes out of my pocket book.

We each had a separate sleeping bag of reindeer skin which was a much more comfortable arrangement than one to hold the three of us. Once inside your bag and toggled up with the flap over head and all, you feel quite comfortably apart from your companions.

Sun 9 Nov   Twenty-fourth Sunday after Trinity. 4 a.m. blowing as hard as ever and continued to do so with the sun out till 2 p.m., when we turned out and cooked a big breakfast, having lain in our bags with nothing to eat for 18 hours. The reason why one is so averse to cooking a meal during a blizzard is that the mere opening of the tent, and going to the sledges to get the food and fill the cooker and get the lamp and so on, brings such an awful lot of snow into the tent that it is less uncomfortable to remain warm in your bag and put up with the hunger.

Wind force 4 to 5 from southwest, but cloud clearing from the Bluff. 4 p.m. wind began to drop, sun bright and clear. Turned into bags again at 5 p.m., after digging out the tent door for the third time today. The sledges are all but buried. Temp. $-2°F.$ to $-7°F.$ during the night. Now plus $3°F.$ All of us very fit and well except Shackle, whose cough seems very troublesome.

Mon 10 Nov   Turned out at 6 a.m. after a disturbed night, during which the depot party caught us up, having marched all night. They camped close by. Two of their party are snowblind, otherwise all well. After breakfast we went off at 8 a.m. leaving them all asleep. And covering some 10 miles without a stop we reached Depot A at 2.30 p.m., the spot where the previous spring sledging trip, consisting of the Captain, Shackleton and the Bo'sun[5] had left a quantity of stores which we now take on south, making up our weights to the full measure again. All the morning we travelled against a head wind, which dropped when we reached the depot and left us a glorious still sunny afternoon to shift our empties, refill our food supplies and repack our sledges. Various jobs, with tea and supper at 9 p.m. and a long cold time for me sketching a very glorious view at several degrees below zero, and then turn in about 11 p.m. Here at the depot, and several times during the afternoon, we were visited by Snow Petrels, the most beautiful of all the Antarctic birds, like snow-white doves. They hovered round us in much curiosity and the dogs were mad to get after them. We turned in hoping that the depot party would soon turn up and enable us to go on in the morning.

Tues 11 Nov   Lay in till noon, as there was wind and drift flying and no sign of Michael's party. Then we had breakfast, and turned in again, read a chapter of Darwin and so on, till at 5 p.m. the depot party arrived and camped close to us and we then had our supper while they had theirs. I then went the round of all Michael's party, four tents, three in each, and had a talk and a laugh, for they were all in very high spirits at beating the 'Farthest south ever reached by man' record of Borchgrevink. Examined them all medically and found them all well and sound and free from any trace of scurvy. They had done a night march of 7 miles on Monday evening and 4 miles today, notwithstanding the drift. All our appetites are already enormous. We turned in about 8.30 p.m. Temperature −2°F. Little wind, but overcast, stratus, not much sunshine.

Wed 12 Nov   Turned out at 6 a.m. Blowing and very thick. But not much snow falling. No land in sight. We all started off together, Michael's party and ours, we having filled up our weights to 2000 lbs, relieving the other party in order that they might travel more up to our pace. We covered 10 miles between 8 a.m. and 7 p.m. taking nearly 2 hours' halt for lunch.
The dogs pulled very well indeed and we had a good going surface, few and small sastrugi, all running southwest. All the forenoon we were in a dense whiteness, no sun, nothing visible, and yet nothing falling apparently, except some fine drift. At noon this cleared and we had an hour of bright sunshine, the rest of the day cold and grey with cirro-stratus. A hard day's work hauling at the head of the dogs half the day and on the sledges themselves the other half. Fine solar

halo with an inverted arch of light above it. Both parties camped tonight farther south than Borchgrevink ever got.

Dailey, the Carpenter, with five men returns to the ship tomorrow. Michael Barne with the other five comes with us for two days more, then we fill up from him to our full weight and go on south alone, while he returns to the ship. Weather perfect. Temperature −2°F.

Thurs 13 Nov   Turned out at 7 a.m. Beautiful sunny morning. Of course we have the sun all day and night, though it is much lower in the south at night than it is in the north at day time. Much time was spent in rearranging stores for us to take on and a large group was taken of the whole party 'father south even than Borchgrevink'. All our flags were flying with the Union Jack. Dailey and 5 men returned to the ship.

From 11 a.m. till 1.30 we covered 5 miles, then we lunched and afterwards managed to get in 3 miles before a sou'wester broke on us which we had been watching as it rapidly came upon us. We had very heavy loads, over 2000 lbs, but the dogs managed very well. Turned in after supper, about 7.30 p.m. when it was blowing a proper blizzard. Temperature hovering about zero, but down to −10°F. during the morning. At present finnesko are the most comfortable boots for travelling. The snow is too cold for leather ski boots. Surface fairly good, some soft patches and some heavy sastrugi deeply undercut.

We crossed a definite long low hill today which quite suddenly hid White Island from us.

Fri 14 Nov   Turned out at 6 a.m. and on march by 8 a.m. And till 8.30 p.m. when we turned in, we had a blinding, shadowless whiteness, the most trying light there is in these regions. Good surface, though soft, and occasional large banks of snowdrift, which one discovered only by walking into and falling over. We did just 10 miles in the day, Michael's party keeping just ahead of us with light loads. We are heavily laden and the dogs pulled well. More than once today the snow crust over some acres subsided to our weight, giving one the weird sensation of a guncotton explosion underground, the rush of air seeming to try and drag one's cap off. The dogs were terrified and cowered. We all camped together about 8.30 p.m. Michael lunched and supped in our tent. Everyone is very fit indeed, except for a few touches of snow blindness.

Sat 15 Nov   At 3 a.m. it was still overcast, nothing to be seen but large soft crystals of snow making deep flocculent snowdrifts. At 8 a.m. we turned out to a beautiful clear sunny day, and the whole morning went in odd jobs on the sledges, as today we fill up from Michael and he goes back to the ship. We took from them all we could carry and started off with terribly heavy loads on wooden runners on a most infernally heavy surface of flocculent snow. After a mile of very heavy work we camped for lunch. One of the bitches we sent home with Michael, as she was doing no work and was miserably cold in a half-shed coat.

It was a gloriously hot day, and after lunch we worked like blazes till we couldn't get the dogs to budge another inch and we have only covered 2½ miles in the whole day. We are glad to have dropped the other parties, so that now we can shove along as fast as we can, though today's work is not very promising. The sun was so hot and the air so drying that we turned our sleeping bags inside out and gave them a chance to dry.

Sun 16 Nov   Twenty-fifth Sunday after Trinity. A perfectly beautiful day, and we turned out at 6 a.m. We first tried the dogs with the total weights and six sledges. They were altogether too much for them on this surface, so we decided to start relay work. We did 2½ miles three times over before lunch, bringing up 3 sledges at a time, then taking the dogs back and bringing up the other three. We did the same after lunch, covering 15 geographical miles in the day and making 5 miles good to the south. Tedious work. We had most brilliant sunshine all day. After camping, there was a fine dog fight in which every one of the team succeeded in joining, as both the dog pickets were dragged out. A white low mist rises over the Barrier surface during the small hours of the night after these sunny cloudless days, and in the mist is nearly always to be seen a white fog-bow facing the sun.

Mon 17 Nov   We started off again about noon and again covered 15 miles to make 5 miles southing during the day. Half an inch of very sticky soft ice crystals covering the surface and making it deplorably heavy for the sledges. They are too heavy, even 3 at a time, for the dogs and we have to pull as well the whole way. The dogs are certainly tiring with the hard work. Sunny morning, overcast afternoon. Turned in about 11.30 p.m. after a long and tiring day.

Tues 18 Nov   Turned out 8.30 a.m. Away at 11. Covered nearly 18 miles today in making about 6 to the south, 2½ by 4 p.m. when we lunched and 3½ by 9 p.m. when we camped for the night. Still the same soft sticky surface. Never cleared all day. Remained calm but misty, a blank white desert. Our appetites simply ravenous and sleep comes without asking. I had a touch of snow glare in my eyes today. Sunburn and frost has made us very sore round the nose and lips.

Wed 19 Nov   Weather as before. Dead calm all day and very misty. Ice crystals falling all day, very thickly tonight. Altogether beautiful weather as the sun breaks through now and again and patches of blue sky overhead. Hard day's work. 15 miles covered to make 5 miles southing. Dogs getting very tired and very slow. We were at it from 11.30 a.m. till 9.30 p.m. and now at 11.15 p.m. we are at last in our sleeping bags.
Surface worse than ever, with a thick coating of loose ice crystals like fine sand. We pray for a wind to sweep it all off and give us a hard surface again. This is wearing us out and the dogs, and yet we cover no ground. And the exertion of driving the poor beasts is something awful. Fine halo with brilliant prismatic parhelia round the sun today. There was a double halo for a while at 4 p.m., the outer one having the more marked prismatic colours and a radius about twice that of the inner. Thick fall of ice crystals again tonight. Have seen no land today, flat Barrier surface all round, very few sastrugi visible. Seriously thinking of giving the dogs a day off to rest.

Thurs 20 Nov   Today we turned out at 10 a.m. having decided to give the dogs an easy day. We made 3⅓ miles southing, covering 10 miles in the day. Perfect weather. Clear sunny blue sky, clouds all along the land line though. The surface today was heavier than ever and inches deep with soft feathery ice crystals. We are all pulling our hardest with the dogs and yet things cannot be made to move faster. I am afraid we shall disappoint the ship in their expectations of a far south record. We are doing so very little now, and all we long for is a heavy wind and better surface. At present we can only worry along till our loads are reduced and hope to cover more ground someday. It is very heavy work for everyone.

Fri 21 Nov   The finest day we have had. Not a cloud in the sky. Very heavy day's work. Surface inches deep with snow crystals, beautiful feathery six limbed stars of considerable size, just the things one sees done in white on a black background in text books. Covered 12 miles today, relay work, making good only 4 miles southing.
We saw new land distinctly today on the southwest horizon, new patches cropping up farther and farther to the south of west. So we altered our course a little from due south to S.S.W. in order to strike

land if possible and leave a depot, making it possible to do bigger distances south with lighter loads and then pick up the depot on our return. Also of course we must map out and sketch this new land which is farther south than Michael Barne's survey trip will reach. Also we hope that nearer land we may find the wind again, and with it a better surface for travelling. The dogs made terribly heavy weather of it today, and the dog driving has become the most exasperating work. Tonight again a fine weather fog and white fogbow and again a fall of snow crystals. Dead calm as usual. Nose and lips cracked and face peeling, very sore. Temperature down to −10°F. Le bon temps viendra.

Sat 22 Nov   Northerly breeze, sufficient to just help us with both sails set. Beautiful day, but threatening to become overcast and windy from the northwest. Turned out at 9 a.m. Made 2½ miles southing before lunch, covering 7½ to do it, pulling hard with the dogs all the way. After lunch we made 2¼ miles, 4¾ to the S.S.W. during the day, and fourteen miles covered to make that. More new land appearing to the southwest, and during the afternoon land appeared right ahead to the S.S.W., which is very satisfactory, as we can now make straight for it and leave a depot, and so reduce our weights considerably. Altogether much more promising than this slow and tedious plod to the south on an ice plain simply to beat a southern record. Now we have new land to survey, and I have the prospect of sketching, and we may find out something too which will explain this extraordinary Great Barrier, as we are so to speak getting to the back of it.
The surface today is very soft and heavy. We tried the dogs with the whole weights today, but they could hardly move them. Very tedious this relay work.

Sun 23 Nov   Thought we were in for a blow last night, but woke to a beautiful day again. It looked as though they were having very heavy weather away in McMurdo Strait[6]. We turned out at 9 a.m. and by covering nearly 16 miles made a little over 5 to the S.S.W. Very heavy work indeed and by lunch time, 6 p.m. we were very leggy after 8 miles in soft snow. After lunch I went on ski and found it a great relief. We have land all along on our right hand now as we go and it extends to right ahead. It is all snow covered and must be very high in places, but there are long gaps in between, where no land is visible. We are all a bit footsore. The other two got a touch of snowglare today. We are making for the land ahead to the S.S.W. to leave a depot with safe land bearings.

Sledging scenes

Mon 24 Nov   Turned out at 9 a.m. Started off at noon trying the dogs again with the whole six sledges, but we couldn't get more than half a mile out of them. So we started relay business again. Managed 2½ miles by lunch time, covering 7½ to do so. One of us now remains with the first load to put up the tent and get the lunch ready, while the other two go back for the second load. Shackle started today and remained with the tent. Same routine after lunch, where 2¼ miles were made good. I remained to put up tent and prepare supper. Dogs very weary indeed and terribly slack and the driving of them has become a perfectly beastly business. However we have approached the land by some 5 miles today and no new land has appeared. The other two have a touch of sunglare today.
Perfect weather. Bright sunshine all day and cold breeze from southwest. It was a great treat being left alone in the tent today, away from the other two and away from the dogs. Joe, one of the *Southern Cross* dogs, was turned out of the team today as a mere hindrance.

Tues 25 Nov   Turned out at 9 a.m. Looked like blowing from the south, but all cleared off again and then turned into a 'white silence' day. Overcast and light southerly airs with an occasional shower of

fine frozen snow dust. Covered 14 to 15 miles today and made nearly
five miles southing. The dogs are dead beat this evening. We go
about a mile an hour, so tomorrow they are all to have a rest. It is all
most disappointing, this inability to cover any distance. We find by
today's sight that we have crossed the 80th. parallel and are now in
80°1½'S.lat. My full day on, the Captain having the morning
camp, and Shackle the afternoon. No land was in sight at any time
today. We had to steer by compass and dial.

Wed 26 Nov   Having decided to give the dogs a day off, we were
not much put out when we got up for breakfast and found it blowing
thick and drifting, so that anyhow we couldn't have done much. We
had a day of rest with the dogs. Read a chapter of Darwin. I did a
good bit of darning and mending and we had two meals. It cleared
towards evening and the new land came out very clear to the S.S.W.
Mount Discovery now looks like a mere mole hill and Erebus and
Terror are clean gone out of sight. I think the apparently flat plain
we are on is really a succession of very gradual rises and falls. Often
today I had the feeling that I was on a hillside of very low gradient,
and when on ski the slip was always in the same direction down the
hill. Yet of course to the eye everything looks a dead level.
Our meals now consist of bacon for breakfast, chopped in small
pieces and fried with pounded biscuit, about a breakfast cup full
each of this. With this, two large cups of tea and a dry biscuit or two.
For lunch we have dry biscuit and two cups of hot Bovril chocolate
with sugar and somatose. For supper we have a thick soup of
pemmican, red ration (pea meal and bacon powder), pounded
biscuit, a soup square (Lazenby's), some powdered cheese boiled up
in water with pepper and salt, making each of us two pannikins, i.e.
about two large cups full. After this we each have a cup full of sweet
hot cocoa boiled with plasmon, and some dry biscuit. These are days
of full meals. We are eating just about as much as we want and
cooking three times a day.

Thurs 27 Nov   Turned out at 9 a.m. to an absolutely cloudless
day and hot sun. It is now 10 p.m. and we are in our bags with the
temperature down to zero, but the sun still hot. Last night I woke up
so hot that I had to throw open my sleeping bag. We came on a mile
this morning with full loads, but the dogs required too much beating
and eventually gave up trying to pull. So we again started relay work
and covered 12 miles in making 4 miles good to the S.S.W. I had the
morning camp and was able to make a sketch of all the new land now
in sight. The dogs worked very badly today, feeling the sun's heat

very much, so we have decided to try night marches in future to see if they will work better.

The coast we are making for is still about 50 miles away and will take us some time to reach at this rate. It looks very beautiful though, all snow covered peaks, bold cliffs and headlands. My right eye got sunglare, feels gritty and the sight blurred, the left eye is all right. Appetites simply immense and *how* we enjoy our food, *and* our sleeping bags, though the joy of breakfast always makes up for necessity of turning out in the morning. We are all very fit and well.

Fri 28 Nov Again a brilliantly sunny day, but with a cold breeze. We turned out just before noon intending to work into a night routine for the dogs. We started off about 2 p.m. and did 2 miles by relays before lunch and after lunch about 2½ more, covering in all over 13 miles the day. I had the tent pitching for the evening, and as it had come over a very thick white fog, depositing hoar frost moss-deep on everything, I had a very cold 2½ hours waiting for the other two with the second load. So cold that I got into my bag. Supper soon warmed us up and we turned in at 1.30 a.m. when I did some darning and wrote up my diary. Temperature down to zero.

The dogs certainly pulled a bit better tonight than they did in the heat of the day, so we shall continue night marches for a bit. We ourselves sleep very warm during the heat of the day. Hunger is having its effect on the dogs as well as ourselves. They are constantly breaking loose, dragging the pickets and going for one another while they are being fed. We had a grand mix up and general fight this evening over the feeding time. Dried codfish, Norwegian *torsk* is what we carry for them.

This fog has gradually come up on us from the south and now hides the sun. We again had a fog-bow this evening.

Sat 29 Nov Turned out about 1 today. Covered 15 miles today making 2½ S.S.W. before lunch and as much after lunch. Very heavy going, as we are certainly now traversing very long and low hills and valleys. We saw a very wonderful exhibition of mock suns, parhelia, perihelia and circles of light in the sky connected with the sun in a blue sky, with showers of ice crystals flying about, just for all the world like spring showers of rain at home as one sees them from the Crippetts. This was such a striking sight that we at once halted and got out the theodolite and took all the angles and elevations and bearings of the various circles and mock suns. It was a wonderful and very beautiful sight but very hard to describe, but I got sketches which I hope will give some idea of it. I had never seen anything like it before. It beat all the halos and parhelia we have ever seen at the ship.

The Captain had the luncheon camping, and Shackle the supper camping. My full day on. The last 2½ miles we nearly lost ourselves, as it came on so thick with a breeze and drift and such a bad light that we couldn't retrace our tracks or see anything at all for a long time and had to go by compass to find the tent. Eventually we found it, but it was a very unpleasant hour. We turned in about 3 a.m. Drifting and blowing with a breeze, force 2, not very promising.

We haven't seen land now for 2 days. All well. Today we began using our dried seal meat.

Sun 30 Nov Advent Sunday. Turned out about 2.30 p.m. having slept nine hours without waking or dreaming. Still white and thick all round and several inches fall of soft feathery ice crystals. Temperature of the air delightfully warm — plus 30°F. Covered 12 miles in the day. Making 2½ before lunch, which we had between 9 and 10 p.m. and 1½ before supper at 3 a.m. Very heavy work for the dogs and for ourselves over this soft surface. The dogs are very done. I had the morning camping and cooked lunch and the whole coast line cleared beautifully, so that I got a sketch of it all at lunch time, i.e. about 10 p.m. Very shortly after, the regular night fog and the white fog-bow turned up and it was all hidden again. We now appear to be much closer in to the land and still more new land has appeared to the south. This is very tedious work though, and the progress terribly slow. The amount of shouting and beating that the dogs want before they will do any work at all is soul sickening.

Mon 1 Dec Turned out shortly before 3 p.m. I had woken in the day so hot that I had to throw my sleeping bag open and open the tent. The sun had baked us inside at midday. Again covered 12 miles in the day, making 2½ before lunch and 1½ after. Very heavy and soft surface. Snow grains falling all day. Nothing but 'white silence' all round us. Dog Bismarck quite done up today refusing to eat. Brownie also very done. We are trying not to overdo it with the dogs and short of doing nothing, we couldn't be doing much less. I had the evening camp duty. We turned in about 5 a.m. Still thick and snowing. Several times again today the snowcrust has 'settled' under our weight with the same startling rush of air that I mentioned before.

We started on a reduced ration today, having taken a little out of each week to make up enough for an extra week out.

Tues 2 Dec   5 a.m. really begins the day, when we are writing up
our diaries in our sleeping bags after a long night's work, preparatory
to turning in for the day. Slept nine hours solid. Turned out at 3
p.m. Covered 13 miles in making 4½ to the southwest. My full day
on. The Captain had the luncheon camping and in preparing our
hot stuff set the tent alight, luckily just as we came up with the
second loads, or the blessed thing would have been burnt.
Providentially I was able to grab the thing the moment the flame
came through to the outside and put it out, so that the only damage
was a hole you could put your head through.
We had a heavy day and turned in late about 6 a.m. The weather
from 5 p.m. till midnight was overcast and we could see nothing.
Then it cleared quickly and the new land all came in sight ahead of
us, more and more new land appearing to the south. One can now
see a lot of detail in it, and it seems to consist of a series of fine bold
mountain ranges with splendid peaks, all snow-clad to the base of
course, but here and there rocky precipices, too steep to hold the
snow, stood out bold and dark. It was a wonderful sight, the pale
blue shadows in the white ranges standing against a greenish sky.
The dogs pulled well until the last lap when the sun was very hot and
they 'threw their 'ands in', and refused to do anything. The only
thing then is to beat them and get them on yard by yard, sickening
work. Many of them are badly chafed by the harness.

Wed 3 Dec   6 a.m. just turned in and writing up diary in my bag.
Bright sunshine. Temperature of the air plus 4°F. taken by a sling
thermometer. Slept 9 hours solid. Turned out 3 p.m. Brilliantly clear
sunny day, blue sky, cold breeze. My morning for camp duty.
Covered 13 or 14 miles today, making 2¾ before lunch and 1½
after. We are now closing the land more apparently. All night we
had perfect weather, clear and blue and the mountain ranges have
looked very beautiful all the time in their pure whiteness and pale
blue shadows. Brownie collapsed again today, preferring to be
dragged to keeping on his feet. He is very weak but works on.
Bismarck off his feed, but certainly improving. The dogs are all very
tired indeed though, owing to the heaviness of the work in this soft
snow. Our position today 80°20'S.lat.

Thurs 4 Dec   6.15 a.m. Just turned in after 14 hours on the
march, including 5 or 6 for camping. The land we are making for is
about 20 miles off and the high range beyond, about 37. We started
later today having slept on till 4 p.m. Very heavy going today,

44

slightly uphill the whole time I believe. The dogs working most deplorably after lunch and we barely made three miles during the day, covering about 9. The weather was perfect, clear blue sky and sunshine and the snow-clad ranges ahead and to our right were a sight worth seeing.

In the early morning about 4 a.m. when we came to turn in, there was a mist and a well marked white fog-bow. On Advent Sunday I made a thorough search in all three of us for any trace of scurvy and I found not a suspicion in any one of us. We find today that we have run through our first can of oil too soon, so we must now knock off our midday hot meal and began a new cold meal of dried seal's liver, sugar and biscuit. Also at our evening meal we knocked off our cocoa to save oil. Not very pleasant but good enough, as we must make our oil last out with our food.

Fri 5 Dec Just turned in, 4 a.m. Turned out again at 4 p.m. Nearly every night now we dream of eating and food. Very hungry always, our allowance being a very bare one. Dreams as a rule of splendid food, ball suppers, sirloins of beef, caldrons full of steaming vegetables. But one spends all one's time shouting at waiters who won't bring one a plate of anything, or else one finds the beef is only ashes when one gets it, or a pot full of honey has been poured out on a sawdusty floor. One very rarely gets a feed in one's sleep, though occasionally one does. For one night I dreamed that I eat the whole of a large cake in the hall at Westal without thinking and was horribly ashamed when I realized it had been put there to go in for drawing room tea, and everyone was asking where the cake was gone. These dreams were very vivid, I remember them now, though it is two months since I dreamed them. One night I dreamed that Sir David Gill[7] at the Cape was examining me in Divinity and I told him I had only just come back from the farthest south journey and was frightfully hungry, so he got in a *huge* roast sirloin of beef and insisted on filling me up to the brim before he examined me.

Glorious day again, without a cloud. New plan today. We did 4 miles straight off with the first half load, then had our cold lunch as we went back for the second half with the dogs. Twelve miles we covered to make this four towards our depot. At a mile and a half from camp I went on ahead on ski and prepared the tent and got the supper under way. My eyes are a bit touched by the sun glare today. On these night marches we have the sun straight in our faces to the south and so strong is the glare that it catches one, notwithstanding all one's care. I have worn snow goggles every day since our start.

Sat 6 Dec   5 a.m. Finished supper and turned into bags. Cloudless blue sky and the snow range a beautiful sight. Breakfast at 4 p.m. and then struck camp, though it looked very threatening. No sooner were we ready to start than a blizzard broke on us. We had hoped to cover a mile or two first, but here it was and we were smothered in snow in no time, so there was nothing to do but pitch tent again and get into our bags. Very damp and uncomfortable. The temp. rose to plus 30°F. and the soft snow melted at once, wherever it fell on our gear, soaking everything.

I washed my face today with a piece of sponge and some of this wet snow. It was black with the smoke of our cooker, except where brand new pieces of red showed up from the sunburn peeling. Very sunburnt, freckled, chapped and frost sore our faces are. My eyes, especially the right one very bad today, running profusely and smarting as though full of hot sand, from sunglare. Used cocaine and zinc solution freely.

This blizzard may do good. Gives the dogs and us a rest, and may harden up the surface. All yesterday I had rheumatic pains in the soles of both feet, which quite went off when the blizzard broke. I find I always get rheumatic pains somewhere before these blizzards, so I know what to expect when we get back to warmer climates.

Sun 7 Dec   Second Sunday in Advent. Midnight. Have just had a cold and meagre lunch of a piece of dried seal liver, a biscuit and a half and 8 lumps of sugar. Lay in our bags while the Skipper read a chapter out of Darwin. We talked and slumbered on till 3 a.m. when we had one hot meal which was excellent, but not enough to fill us. We have ever increasing appetites on these low rations and never feel filled by a meal. Nothing to be seen outside — blowing and snowing very hard.

Turned out at 4 p.m. as the wind was dropping and the sun appeared. Had breakfast, dug out the dogs and sledges and started off. Very heavy work again. By covering 12 miles we made 4 towards the land. Dogs very weak and apparently getting rapidly weaker. I think the fish food is at fault. Some of them have a sort of dysentery, one dog Snatcher passing a lot of blood and very weak indeed. Very muggy tiring day. Eyes still very troublesome.

Mon 8 Dec   7 a.m. Just turned in after supper. Overcast sky. Warm temperature, but nothing visible anywhere. Turned out at 4 p.m. to a clear sunshiny day, beautiful and bright, but a terribly soft surface and high temperature. No breeze. We did all we could for two or three hours and managed to drive the dogs over two miles.

Then we had perforce to give it up and leaving Shackle to pitch camp, the Captain and I went back for the second half of the loads, eating our cold lunch on the way. There seemed no possibility of persuading the dogs to pull these, so we tried tempting them on with 2 or 3 fish.

They were all exceedingly hungry and mad to get the fish and by short stages we got them over the two miles. But so weak have they become that though there was barely 500 lbs, they could only manage a few yards at the pace they would have raced away with three times as much, four times as much, at starting. This is but a full load for five dogs and we have eighteen, and they are beaten by it! Evidently they are weakening very quickly and I think the dried fish must be at the bottom of it. No less than four of them are now passing blood in the motions.

Tues 9 Dec   2 a.m. Just turned in. Warm sunny morning. Beautiful range of rugged snow-covered mountains before us, with a long rounded snow hill at the foot. Did some darning. Woke up at 10 a.m. and as it was very hot in the tent, I sat outside in the hot sun and made a sketch of the panorama of new land before us. The heat was intense and there was no breeze. So one could easily sketch in bare hands and they got sun scorched. Turned in again when I had finished and slept till 4 p.m. when we had breakfast. Broiling hot sun all day. The dogs managed 2½ miles with half loads, but Snatcher died in the night and on opening him I found he had died of an acute peritonitis.

The snow today was soft, ankle deep, and the work very hot and heavy. No breeze except an occasional puff from the north. No clouds, the sun scorching down on us the whole march. We turned in at 4 a.m., no one having had a rest in camp today, as all three are needed to haul and help the dogs.

Wed 10 Dec   4 a.m. Just turned in after a very hot march. Wonderful weather, certainly continental, cloudless and sunny without a breath of wind. Though we tied the entrance of the tent and the ventilator open and lay in our bags with the flaps all open, we were so hot we slept but little. We turned out about 4 p.m. and managed to make 3½ miles towards the land, covering 10½ to do it, I had the camp shift and spent the time on a sketch. For three weeks and more now we have been at this heartbreaking double work, covering every mile three times over. This morning to our surprise we were visited by a skua gull, one of McCormick's *Megalestris*. He must have scented the dog I cut up at our last camp. At the end of our night's work a southerly breeze sprang up, making it much colder, too cold even to sketch in gloves. The weather still remained clear and sunny.

Thurs 11 Dec   5 a.m. Just turned in. The skua is still with us. Turned out at 4 p.m. after another very hot night in the tent. We started seal meat with our breakfast, as well as having it for lunch today. Very heavy day's work. Covered nine miles and made 3 good towards the land. Soft and heavy snow, the dogs constantly stopping all together and refusing to pull on. The Captain had the camp shift. The dogs seem a trifle better in health for the pieces of their companion, whom I cut up and distributed among them. Not one refused to eat it, indeed most of them neglected their fish for it. There was no hesitation. 'Dog don't eat dog' certainly doesn't hold down here, any more than does Ruskin's aphorism in *Modern Painters* that 'A fool always wants to shorten space and time; a wise man wants to lengthen both'. We must be awful fools at that rate, for our one desire is to shorten the space between us and the land. Perhaps Ruskin would agree that we are awful fools to be here at all, though I think if he saw these new mountain ranges he might think perhaps it was worth it.

The snow is soft and one sinks in at every step, making the walking very fatiguing. Supper, sleeping bag and breakfast are joys worth living for under these conditions. Only our appetites have clean outgrown our daily rations and we are always ravenously hungry, before meals painfully so for an hour or two.

Fri 12 Dec   Overcast with a good stiff breeze from the south, so we are no longer troubled with the heat of the sun, which takes the skin off our faces and soaks us through with perspiration. 6 a.m. Just turned in to our bags. Turned out about 4 p.m. after a good night, with the usual dreams of food. Waking at midday I sat up and wrote a line to Ory, then to sleep again. It was again very warm in the tent. We made 3 miles in the morning with half the loads and while Shackle took camp duty, the Captain and I brought up the other half in the afternoon, really the small hours of the morning. Nine miles covered today and 3 miles nearer the land. *Some* day we hope to get there and drop some of our load, making a depot we can pick up on the way home. Then we push on south, I hope, at a faster rate with lighter loads.

The dogs take it into their heads to pull or not to pull in the most unaccountable manner, generally the latter. The whole of the first stage of our march today they were awful and yet over the same

48

ground exactly, in the second stage, they pulled steadily the whole way, with very little shouting or beating. The whole reason may have been that the snow was 5° lower during the second lap, being 15° instead of 20°F. Otherwise the weather was the same, and I think very few degrees difference in the temperature of the snow may possibly make all the difference between a good and a bad surface for the German silver runners that we have had on our sledges all along. We did give wooden runners a trial at Depot A last month and decided against them, but I am afraid the trial was not a fair one. We are now feeding 8 or 9 of the best dogs with dog flesh and fish and they seem to be picking up a bit. Their appetites are ravenous, but so are ours. We talk a great deal about food — what we would like to have and will have when we first get back. We are really doing very heavy work on very low rations, but are all very fit. I think a lot on the march of the hot summer days at the Crippetts when one could go down into the cool dairy and find unlimited fresh creamy thick milk and a large common 'destroyer' cake. Fresh milk — how I long for a big jug of it!

Sat 13 Dec   6 a.m. and we have just turned in. Sky all overcast with low stratus cloud over the mountain tops. Turned out at 4 p.m. Weather still overcast. After shouting ourselves hoarse for 2 miles we couldn't make the dogs move more than a yard or two without stopping. Terribly heavy soft snow into which both we, dogs and sledges all sink at every step. It clogged heavily under our ski too when we tried them.
As we were now at a spot where we had good enough land bearings to find it again, we decided to leave here a depot of dog food and all the gear we could possibly spare. I had camp duty in the afternoon and several jobs to do, including the cutting up of a dog. We had a long discussion upon what we could leave here and decided to drop in all some 320 lbs here, and further in, when we get actually to the rock, we can leave 300 lbs more.

Sun 14 Dec   Third Sunday in Advent. Now 5 a.m. and we are in our bags. A good stiff southwest breeze blowing. Just finished sewing up a large flag of burberry to mark the depot. Also doctored both the others' eyes. It feels a good deal colder tonight, though the temperature is up to plus 27°F. We slept with the tent door open. Turned out about 4 p.m. to our special Sunday breakfast — NAO ration, tea and biscuit. Spent some hours making a depot of spare gear and dogfish, then started off again towards the land. We hadn't gone ¾ of a mile when we were held up by a ridge of stupendous ice pressure, with crevasses and hummocks of enormous depth and size. We camped at the edge of this and lunched, and then having roped ourselves together, we attempted an investigation of this pressure ridge, but the light was exceedingly bad and we managed nothing, except that some photos were taken.

It was a wonderful sight, a chaos of ice masses jumbled up in crevasses of 40, 50, 60 feet deep, the valleys some hundreds of feet across full of tumbled blocks and frozen pools of water. There were multitudinous icicles formed by the sun's heat on these ice masses, pinnacles and towers and weathered stratified ice cliffs. We came back, finding it impossible to cross this line of crevasses, had supper and then turned in. We decided to depot our own return journey provisions close to where we left the dog fish and spare gear.

Mon 15 Dec   3.30 a.m. Just turned in, supper finished. Warm and overcast with low stratus. Woke in the evening to an absolutely clear blue sky and a still and blazing hot day. Intense heat. Everything in the tent wet from the melted snow. After breakfast we again went in among the crevasses and took a number of photos, as the light was much better. There was a long line of fog or frost smoke along the actual coast line farther in, which I took to mean water, either another line of crevasses masking an actual tide crack or some water exposed at the bottom, or else another line of crevasses in which there was more melting from the sun's action on the blown rock dust off the mountains, which would be there in greater quantities than here some ten miles farther out. The pools of water in the crevasses here are I think undoubtedly formed in that way.

We then returned to our depot, which we call Depot B and left there food for 4 weeks, which is to take us back from here to Depot A. We carry on from here provision for 4 weeks and 5 days, giving us 2 weeks more now to go southward and 2 weeks back here, and a few days to spare for weather and for finding our depot, which in thick weather might be a difficult business. We had our lunch here and got very cold arranging foodbags, and biscuit tanks, as the sky became overcast and a breeze sprang up from the S.

At 11 p.m. we started off south at last carrying our whole load at once and we made 2 miles good by 1.30. It *was* a relief to at last give up relay work and travel with the knowledge that we hadn't to cover the same ground three times over. Turned in after a good supper at 4 a.m., having travelled in a nasty drift with the wind dead ahead for over 2 hours. There was however a beautiful double rainbow halo round the sun, with parhelia and inverted arcs whenever the sun broke through the driving ice crystal showers. Killed another dog tonight.

Tues 16 Dec   Dad's birthday. Many good wishes from the far south. 4.30 a.m. Just turned in to our bags, much the best place to be in during these icy drift storms. We turned out about 3 p.m. after a long night of sleep with the tent door tied open. Blazing hot sun again, so that we have to bury our bacon and fat foods in snow on camping, or they melt in their bags. We then made 5½ miles good to the south between 6.30 p.m. and midnight, all three of us pulling on ski. Then lunch as we stood and the Captain took a sight, and after lunch we made up our day's march to 7 miles by 2 a.m. The view at starting was grand, but the line of frost smoke along the coast gradually rose and spread out towards us till only the mountain tops were visible above it.

We still wear finnesko, the snow being much too cold for ski boots unless one is always on ski, which we cannot be at present. The dogs seem improving on dog flesh diet, but work in a very capricious manner. Nine of the best dogs get dog flesh. The rest still living on fish are getting daily thinner and weaker, especially in the hindquarters. Shackle's eyes are still very bad. The surface we are travelling over is very soft, though smooth as paper with minute rippling sastrugi, and the sun is thrown up from it like smooth water. But there is a crust through which the dogs break and this hurts their feet, so that they shirk it and consequently lose all force in their pull, and are always inclined to stop.

Wed 17 Dec   4 a.m. and a well marked fog-bow of white light. Just turned into our bags after camping and supper. I act as butcher every night now, killing a dog when necessary and cutting him up to feed the others. Slept well and woke at 11.20 when all the fog was gone and it was a brilliantly clear hot sunny day. No breeze. So I left my bag and the tent and sat sketching on the sledges for nearly three hours, when the others turned out and we had breakfast. We spent an hour upon odd jobs and started away south or a trifle east of south towards the land farthest visible.

I was leading, but after 3 hours got such a violent attack of sunglare in my eyes that I could see nothing. Yet I had worn grey glasses all the time sketching, and grey glasses as well as leather snow goggles on the march. I put in some drops and for the remaining five hours pulled behind with both eyes blindfold. They were very painful and streaming with water. By the time we camped they were better. We had lovely weather and the dogs pulled fairly well, making in all some 8 miles during the day. We passed some very splendid cliffs of rock and ice and glaciated land. The sun's heat was intense today, scorching our faces and hands. Noses constantly skinning, lips very painful, swollen and raw.

Thurs 18 Dec Again the sweet job of cutting up the dog, then supper and now in our bags about 2 a.m. Still clear blue sky. Turned out at 11 a.m. Overcast sky, low stratus over all the mountains. Dogs pulled steadily and the surface remained fair till late in the evening, when it became very soft with deep ice crystals. We all go on ski now, pulling, leading and driving. It tires one much more evenly all over, there is no special strain anywhere. The Captain's eyes have been bad today, but he makes little of it. We lunched on a halt, standing for ten minutes and camped eventually for the night at about 9 p.m. Very chilly this evening. Raw, cold and grey, no sunshine. It is now 11 p.m. and we are in our bags, having taken to day marching again, as we don't like missing the coastline in these regular nightly fogs. Butcher's work again as usual, my duty on camping while the others pitch tent. Again vivid dreams of a splendid spread of food but got nothing this time. Our hunger is very excessive.

Fri 19 Dec Turned out soon after 9 this morning. Overcast sky, but a very dazzling light. We all went on ski. Covered 5 miles before lunch, pitched tent and sat in our bags for an hour to give the dogs a rest as they seem very tired out. After lunch we covered another 2¾ miles. The dogs now seem to be failing rapidly in strength, I think chiefly from a form of scurvy from the fish we are feeding them on. We have decided now to feed up 8 or 9 of the best dogs on the others, and when these dogs have eaten each other up we must pull our own victuals and gear. We turned in at about 10 p.m. Very warm evening, plus 27°F. by the sling thermometer, which gives the true temperature of the air, cutting out all solar radiation. Tent door tied open again. Overcast still, low stratus. The surface today as smooth as paper, but very soft under a thin crust in most places, letting one's ski down an inch or more.

Sat 20 Dec Woke in the night and lay awake for two hours from sheer hunger. Got my breakfast allowance of biscuit off the sledge and slept at once. Turned out at 8 a.m. Made 6 miles before lunch, camped for lunch and did 2 miles more in the afternoon, as we got into terribly heavy snow at plus 27°F. in which the sledges ran so heavily that the dogs could hardly pull them.
The sky was overcast all day, and there were occasional flakes of snow falling. Only the lowest rock patches were visible on the coast. One of the dogs dropped dead in harness today. Several others look as though they would like to. We are all very fit and well but thin, and our appetites ravenous. They tell me I am even more gaunt than before, certainly thinner, but very well. We all take turns at cooking the meals.

Sun 21 Dec   Fourth Sunday in Advent. I now save half a biscuit from supper to eat when I wake at night, otherwise I simply can't sleep again. I have never experienced such craving for more food before. We turned out about 7 a.m. and after breakfast made 2¼ miles when the dogs became so utterly rotten that we camped. It was very hot and close and we decided to start again at night time. Soft snowflakes falling, no sun, but low stratus. We eat our cold lunch and then slumbered and did odd jobs from 1 till 7 p.m. when we had some hot Bovril chocolate with somatose, and made another start. The dogs were worse than ever and after a mile and a half we had to give it up. They simply wouldn't attempt to pull and seem as weak as kittens. We then tried some experiments — uncoupled all the dogs and took off all the food we were carrying for them, leaving only our own kit and food for the month. The surface was so heavy that we three could hardly budge it at all. We camped and had supper and turned in for a short night, intending to get back to the day routine again tomorrow.
The night was as raw and cold as the day had been oppressively hot, up to plus 28°F. by the sling thermometer. Did my butcher's job and turned in.

Mon 22 Dec   1.30 a.m. Just turned in. Cold northerly breeze, soft feathery snowflakes falling. Very little visible, only the lowest rock patches. Turned out 9 a.m. to a gloriously bright sunny morning, the sky having cleared all round during the night. We made 4 miles to the south and then camped for 2 hours, during which I was sketching. We then did 2 miles more, 6 in all, very fair considering the condition the dogs are in and the very heavy snow we are travelling over now. We camped about 9 and turned in about 11. Conversation runs constantly on food. We are all so hungry. The sunshine today was a blessing, though our lips and noses and faces altogether are very sore. Our position today is 81°22'S. lat. The coast line we are exploring is certainly a very grand sight and full of interest and new sights appear every day. This afternoon a heavy white fog came up and blotted everything out. Cleared in the evening, so I sketched, but again came over us later on with a beautiful white fog-bow.

Tues 23 Dec   Turned out 9 a.m. bright warm and sunny. Got a sketch before we started. Our time here is 11 hours 12 mins. ahead of Greenwich time as we are in 163°23'E.long. The dogs pulled steadily all the morning and afternoon. We camped for lunch about 5 p.m. having made 5 miles the forenoon and then made 3 more before camping at 10.30 p.m. for the night. Now just turned in about 2 a.m. Three hours is about the usual time for camping, cooking, having supper, turning in and writing up diary. This has been a more hungry day than ever, as we have had to reduce our biscuit and seal meat for lunch and breakfast to make it run its appointed time, and we don't want to turn back a day earlier than necessary. All in good spirits. A very cold southerly breeze to face all the evening march, as low as plus 8°F. which we feel badly on our sunburnt lips and faces. Got a satisfactory sketch again today.

Wed 24 Dec   Christmas Eve. 2 a.m. In our bags writing up diaries or talking of food, letters and the relief ship. Turned out at 9.30 a.m. to a bright sunny morning, nice and warm. We had several jobs to do. We discarded and left the large sledge which carried all the dogs' fish and cut in two our own provision tank to take also the remains of the fish and carry the dog flesh. This meant a good bit of sewing as they are canvas tanks. Shackle and I did this, while the Captain took a round of angles and a sight, which put us at 81°33½'S.lat. We started away about 1 p.m. and made 5 miles by 6 p.m. Camped for lunch and then did 3 more by 9 p.m. when we camped for the night.
As a result of today's medical examination I told the Captain that both he and Shackleton had suspicious looking gums, though hardly enough to swear to scurvy in them. No sketching today, very hazy light and excessive mirage. Surface rather better and dogs pulling better. My eyes touched by sun glare very painful during the night.

Thurs 25 Dec   Christmas day. Just gone midnight. A Merry Christmas to all at home. We are in our bags writing up diaries, looking forward to full meals for once. Turned out at 9 a.m. to a glorious Christmas of blazing sunshine. We were cooked by it all day, except while we were cooking. We had three hot meals! I read Holy Communion and various other things in my bag before we turned out. Our meals must be given in detail as they were very exceptionally good today. I cooked the breakfast. We had tea, extra strong and sweet. (Milk of course we haven't had since we left the ship.) Biscuit, and a pannikin full of biscuit crumbs, bacon and seal liver fried up in pemmican. To top up we each had a spoonful of blackberry jam from a tin we brought specially for this day, our only tin. After breakfast we grouped ourselves in front of the camp and let off the camera by a string, flying all our flags and the Union Jack. We then did a good 6 miles' march and camped for lunch in great heat. We had a brew of Bovril chocolate and plasmon, biscuit and more blackberry jam. The Captain took a sight and I made a sketch, but my left eye is useless. We then did 4 miles and camped for the night at 8.30 p.m., having covered 10 miles in from 6 to 7 hours, a great improvement. A snow peak abreast of our Christmas camp is to be named Christmas Height. Shackle cooked our supper. We had three NAO rations, with biscuit and a tomato soup square from our 'Hoosh McGoo'. Then a very small plum pudding, the size of a cricket ball, with biscuit and the remains of the blackberry jam and two pannikins of cocoa with plasmon. We meant to have had some brandy alight on the plum pudding, but all our brandy has turned black in its tin for some reason, so we left it alone. We enjoyed our Christmas, though so far from home.

Fri 26 Dec   Woke up at 5 a.m. and as the left eye is still uncomfortable, made a sketch, using the right eye only. About 10 a.m. we started off and made nearly 5 miles, when my left eye got so intensely painful and watered so profusely that I could see nothing and could hardly stand the pain. I cocainized it repeatedly on the march, but the effect didn't last for more than a few minutes. For two days too I have had this eye blind-fold for a trifling grittiness and now it came to this, while the right eye, which I had been using freely was perfectly well. The Captain decided we should camp for lunch and the pain got worse and worse. I never had such pain in the eye before, and all the afternoon it was all I could do to lie still in my sleeping bag, dropping in cocaine from time to time. We tried ice, and zinc solution as well. After supper I tried hard to sleep, but after two hours of misery I gave myself a dose of morphia and then slept soundly the whole night and woke up practically well.

Sat 27 Dec   Turned out at 7 a.m. Again a bright sunny day and no wind. I lay in my bag with my eyes bandaged while all the cooking was done, for fear of starting off again. The Captain and Shackle did everything for me. Nothing could have been nicer than the way I was treated. We started off at 10 a.m. and without camping for lunch, made a good march of 10 miles by 7 p.m., through a long day with a scorching sun. We then camped for the night.
From start to finish today I went blindfold both eyes, pulling on ski. Luckily the surface was smooth and I only fell twice. I had the strangest thoughts or day dreams as I went along, all suggested by the intense heat of the sun I think. Sometimes I was in beech woods, sometimes in fir woods, sometimes in the Birdlip woods, all sorts of places connected in my mind with a hot sun. And the swish-swish of the ski was as though one's feet were brushing through dead leaves, or cranberry undergrowth or heather or juicy bluebells. One could

55

almost see them and smell them. It was delightful. I had no pain in the eyes all day, a trifling headache. Towards evening we came in sight of a splendid new range of mountains still farther to the south.

Sun 28 Dec   Innocents' Day. Turned out at 7 a.m. to a cloudless sunny day with a fierce sun and no wind. Variable breezes, chiefly northerly. Marched again the whole day with both eyes blindfold, as I want to be able to sketch this new range, the biggest we have yet met with and farthest south we shall see in all probability. We covered about 6 miles today on ski and camped early at 5 p.m. to take photos and make sketches and the Captain to take a round of angles. Our noon sight today puts us on 82°10'S.lat. and though we shall not have done a good record towards the South Pole, we have had the unlooked for, hardly expected, interest of a long new coast line with very gigantic mountain ranges to survey and sketch, a thing which to my mind has made a far more interesting journey of this than if we had travelled due south on a snow plain, for so many hundred miles and back again. My left eye is still quite useless and the sight all blurred, but I got a sketch of the whole grand sight before us with one eye. Now it is 9.30 p.m. and we are just turned in to our bags.

Mon 29 Dec   Woke up to find a stiff southerly breeze blowing into our open tent door, with a lot of ground drift. We had intended making an excursion today to investigate the immense ice pressures that were apparent to us yesterday, at the mouth of the strait which we faced in our camp. Though the temperature was plus 15°F., it felt very cold indeed. We breakfasted at 8 a.m. and decided to await the weather, as it was not safe to leave camp at present. About 2 p.m. we had lunch and at the same time there was the most wonderful exhibition of mock suns, solar halos, par- and peri-helia that I have ever seen, and more complex than anything I have ever seen drawn or described. No less than nine mock suns were visible at once and arcs of fourteen or more different circles, some of brilliant white light against a deep blue sky, others of brilliant rainbow. The cause of all this was a very low drift of ice crystals over our heads, but not thick enough to obscure the blue sky and the sunshine. Dogs, sledges and tent are half buried in snowdrift. Stripes was cut up for the team today, he having dropped on the last march.

This evening the land is all hidden, so we are thankful we got angles, photos and sketches yesterday. A thick white fog has now come down on everything round us, the wind has dropped, and the temperature has risen again to plus 23. This day of rest was certainly needed both

by dogs and ourselves, but it would have been nice to have got
another sketch from here in a better light. The mountain range
facing us now has peaks ranging from 10,000 to 13,000 feet. The
highest, which is to be named Mount [Markham], is a thousand feet
higher than Mount Erebus. We are rather surprised that we haven't
come across any volcanoes either active or asleep. All our mountains
are in all probability granitic, simply a continuation southwards of
the already known line of mountains of South Victoria Land and the
general character is the same. Mount Melbourne, Mounts Erebus
and Terror and Mount Discovery are apparently the only three
outbursts between Cape Adare and 82°17′S.lat. which is the farthest
we reached.

Tues 30 Dec   Turned out at 6 a.m. Eyes practically well, except
for blurred sight in the left one. Took on cooking again. Twenty
minutes exactly is the quickest record for breakfast. We time every
meal now from the moment the lamp is lit to the moment it is put
out, competition making us all very smart about the job, and so
saving oil, of which we have a minimum allowance. Twenty minutes
suffices to melt and boil snow enough for a good 4 pints of tea and a
fry of broken up biscuit, bacon, pemmican and seal meat. At 10
a.m. we started off this morning in a dense fog, so that nothing was
to be seen. We got in among a lot of high ridges of ice pressure,
which made us steer out again and after covering some 4 miles, we
pitched our southernmost camp and waited, hoping that the fog
would lift and allow us to look right up the strait we faced.[8]
We knew that we were now in a position to look right up if only the
weather would clear, and we wanted very much to know if we had
struck an open strait or only an inlet, with land visible at the head of
it. The Captain and I went for a ski run this afternoon to the south,
but saw nothing and were compelled to return when we had gone a
mile or two, as we were afraid of losing our camp, the weather was so
thick.
Tomorrow, whatever happens, we must turn north again, our
farthest southern point being 82°17′S. lat. and our farthest south
land charted to 83°S. Just about 300 miles of new coast line we have
got now, from the ship's position in 77°48′S. to 83°S.

Wed 31 Dec   Turned out at 6 a.m. NAO ration breakfast. Sun
coming out and clouds rapidly clearing. We started off and made 4
or 5 miles by 2 p.m., watching the strait all the while to see if the
head of it would clear. The mountains were perfectly beautiful today
in the sunshine. As far as we could see, and apparently it was a blue

horizon line, there was no land blocking the strait, so we must suppose this southernmost high land of 13,000 ft to be insular perhaps. The strait was about 20 miles or more across and ran in due west, I should say. But all these details will appear when the Captain's map has been made out. He took a sort of running survey of the whole coast line. All the morning we were crossing very immense pressure ridges radiating from the cape[9] which formed the northern boundary of the strait. They were cut up in all directions by immense crevasses which were all filled in and bridged over with compacted snow. Sometimes from edge to edge the crevasse would measure 50 to 60 ft and the whole train of sledges, dogs and all would be on the bridge at once. Only at the edges was there a risk of going through and some of the narrow crevasses too let one down suddenly. At 2 p.m. we camped and after lunch, with some food in our pockets, we started off on ski to try and reach bare rock and bring back some specimens for the geologist. We also wanted to know whether there was anything in the way of a tide crack on shore or not, as this would practically decide whether the Barrier was afloat or aground. We ran in some 4 miles over smooth rounded pressure ridges, hills and vales and then were brought up by a perfect chaos of pressure and crevasses. We roped ourselves together, took off and left our ski at the edge and commenced to try and cross what appeared to be a gigantic tide crack, extending about a mile and a half across to the snow slopes that came off the land. Once across this, we could get our rock specimens which appeared quite close, but we had more than we could manage before us.

We started by going down steps cut in an ice slope, then by continual winding from side to side we made our way gradually across what at first looked like impassable crevasses, but in places they were filled in, though 50 to 80 ft deep and blue ice, and in places they were bridged over. But after a while we were faced by more and more precarious bridges, and they got narrower and fewer, and were constantly giving way as we crossed them one by one on the rope. We never unroped the whole time, as there were crevasses everywhere and not a sign of some of them, till one of us went in and saw blue depths below to any extent you like. Shackleton was tied up in the middle, and the Captain and I at each end. Sometimes he led, sometimes I, if he came to an impasse and we had to go back.

As we got deeper and deeper in among this chaos of ice, the travelling became more and more difficult, and the ice all more recently broken up, so that no snow bridges had formed and we were faced by crevasses, ten, twenty, and thirty feet across, with sheer cliff ice sides to a depth of 50 or 80 ft. Unknown depths sometimes, because the bottom seemed a jumble of ice and snow and frozen pools of water and great screens of immense icicles. A very beautiful sight indeed, but an element of uncertainty about it, as one was always expecting to see someone drop in a hole, and while keeping your rope taut in case that happened, you would suddenly drop in a hole yourself. We tried hard to cross all this and reach the rock, but after covering a mile or more of it we came to impassable crevasses, and then saw that the land snow slope ended in a sheer ice cliff, a true ice foot, of some 40 ft which decided us to retrace our steps, as even if we reached it, this ice foot would prevent our reaching rock. The sun's heat was intense and not a breath of wind stirring. The heat has a very great power on these ice masses, as is evidenced by the immense icicles and frozen pools of what has been water. The colours, all shades of blue and pale green and shimmering lights were to be seen among these crevasses. The prismatic colours of the ice crystals were wonderful too today, forming what looked literally like a carpet of snow, glittering with gems of every conceivable colour, crimson, blue, violet, yellow, green and orange, and of a brilliance that would put any jewel in the shade. Our supper got upset in the tent sad to say, and we are so short of food that we scraped it all up off the floor cloth and cooked it up again. It was a soup so didn't suffer much. Another dog died today from sheer weakness.

Thurs 1 Jan New Year's Day. Best wishes to all at home, and the best of good luck to all of us. We turned out at 8 a.m. to breakfast. Fine hot sunny day again, slightly overcast by cirro-stratus. Got away by 11 a.m. and made 4½ miles by 2.30. Camped for an hour for lunch and then made 4 miles more by 7.30, when we camped for supper. 8½ miles in the day, homewards now. We had sail set all the afternoon with a fair breeze. The dogs are terribly weak and of very little use. Another dog dropped on the march today, too weak even to walk. We put him on the sledge till evening, and when we had fed them and gone into the tent, he was killed by his neighbour in the trace, for his food. Our tent floor cloth, a large square of Willesden canvas, we use as a sail and it makes an excellent one. We are now making a course for Depot B, which we *must* reach before January 17th. There we pick up a month's food and then make a course for Depot A. At Depot A we pick up a week's food, which takes us back to the ship. This depot system has risks, and nearly brought disaster to Peary[10] a short time ago, but I think we are fairly safe unless we get continuous thick weather, which would make things very awkward. What we have to consider is that we shall soon have no dogs at all and shall have to pull all our food and gear ourselves. And we don't know anything about the snow surface of the Barrier during summer. It may be quite different to what it was on the way south. One *must* leave a margin for heavy surfaces, bad travelling, and weather, difficulty in picking up depots, and of course the possibility of one of us breaking down. We have been making outwards from the coast today and find the surface improving as we do so. Close in shore it is apparently windless, and there is evidently a tremendous precipitate every night during the summer of fog-crystals, which lie inches deep, feet deep in places, forming a smooth, soft, crustless surface of flocculent snow.

Fri 2 Jan　　Overcast, breezy and chilly. Turned out at 6.30 a.m. Made some alteration in the sledges and discarded the smallest of the three with sundry odds and ends that we can dispense with. Got away about 9 a.m. and made 5 miles N.N.W. by 1.30, when we camped for an hour. After lunch marched for 3 hours and made 8 miles for the day. We were able to set sail to a fair breeze this morning. The surface has been heavy all day, and the air very close, the sun's heat coming strongly through a stratus of low cloud. We gave up pulling on ski today as we find we can make better way on foot, though it is very heavy going. Sun out this evening, fine and hot, no breeze. Another dog dropped today, two others very near it. We have now 10 dogs left out of the original 19. We are now camped opposite the high red cliffs, in sight of Christmas Height.

Sat 3 Jan　　Turned out about 6.30 a.m. Slightly overcast with high cirrus, but blue sky and sunny, a little breeze. I am now permanent breakfast cook. We got away about 8.30 a.m. and covered 5 ½ miles by 1 o'clock, when we pitched tent for lunch. I sat outside on my bag and sketched some magnificent red cliffs about 10 miles off, which we passed at 15 miles on the way south. They are very striking and of immense height — 2000 ft of sheer rock cliff. After lunch we covered 4 miles making 9½ for the day, thanks to a fine breeze which carried our sledges along at a splendid rate, over a surface of soft ice crystals into which we and the dogs sink pretty deep, but over which the sledges run very smoothly and easily when the sun has been on it for a while. We are all in finnesko and find it pretty heavy hauling, but steady work which makes it none so bad. The dogs are rapidly giving up. Two more dropped in their traces today. It has been a grilling day and I have been wet through all day from free perspiration. For 64 days now we haven't had our clothes off, and only once have I scrubbed my face. The sun and wind and frost make our faces too sore for luxuries of that sort; we have too little oil to melt any snow for washing.

Sun 4 Jan   Epiphany Sunday. Good juicy brown beef dripping is one thing I long for, and a large jugful of fresh creamy milk in Crippetts dairy. Killed another dog today as he was too weak to walk. We turned out at 6 a.m., had breakfast and were on the march by 8.30 a.m. And though the surface was very heavy with ice crystals, soft and deep and smooth, there being no sun to glaze the surface, we did 4½ miles by lunch time, when the Captain took a sight, but it was too overcast all over the land for me to sketch. We had an hour's rest and then made 3½ miles more in the afternoon. We have now only 8 dogs and they are good for no work at all. We camped at 4.30 p.m. when the sky cleared over the land, but a cold breeze from the north made sketching impossible. We are all now pulling on foot in finnesko all day, heavy work for 7 hours or more, soft ice crystals with no crust. The sledges go very heavily when there is no sun, but run easily as soon as the sun comes out. I think much on the march of our return to the ship, when we shall I hope, find all our letters waiting for us. Le bon temps viendra.

Mon 5 Jan   Turned out 6.30 a.m. Fine sunny morning with northerly breeze and a little cirrus about. Made 4¼ miles in the morning, lunching at one. I sketched outside on my bag as the sun was warm and the breeze had dropped. Our lunch is a meagre meal — a biscuit and a half, eight lumps of sugar and a piece of sealmeat. One can hold it all in one hand, but it serves to carry us on for three hours' hauling in the afternoon. During this we covered 3¾ miles today in very deep soft ice crystals. We can now pull the sledges ourselves, so we have given up driving the dogs. They are doing no work at all.
We had yesterday the most beautiful cloud colouring round the sun, and again today much the same. There were cirrus clouds on a blue sky round the sun and these were beautifully edged with a vivid scarlet — a real vermilion, which ran into orange, yellow and pure violet on white cirrus clouds. The breeze today continued northerly, but the sun on the snow crystals made the sledges run very easily on the whole. One sweats very freely all day long. We have got into a routine now which balances our work and our food supply to a nicety, keeping us always hungry. We work for 4 hours between breakfast and lunch and for 3 hours between lunch and supper and we get about 10 hours' sleep. The rest of the time is spent in camping duties.

Tues 6 Jan   Turned out about 6 a.m. No breeze, intensely warm, close, and an overcast sky. We made 5½ miles before lunch in 4 hours over an excellent wet snow surface. Good for the sledges, heavy under foot. We got a southerly breeze and set sail, and as the sledges went faster than the dogs, we tied them all on behind. Snow soon began to fall and the temperature rose, till as last the sling thermometer showed a temperature of plus 34°F., two degrees above freezing point! Almost a record high temperature, I believe, for the Antarctic. Every flake was half water as it fell and everything on the sledges got soaked, as though with heavy rain. This persisted all the afternoon, land, sun etc. all being lost to sight, till 7 p.m. We camped at 5 p.m.
Another dog dropped today and was used as food for the others. Just before lunch today we crossed our old tracks, made as we were going south. We have about 62 miles more to cover before we reach Depot B. The dogs are now only a hindrance. We covered 8 miles today and camped in very wet snow, and our bags and gear in a very wet state too.

Wed 7 Jan   Turned out at 6 a.m. Bright warm and sunny morning. Started off with a light head wind from the northwest. Cast all the remaining dogs adrift and pulled the sledges ourselves on foot. Good surface. Made 5½ miles by 12.30. Camped for lunch. All the land partly obscured by low stratus. After lunch made up the 10 miles and camped at 5 p.m. Very hot sun, but more or less overcast sky. Very free perspiration, wet through all day. Washed our feet and hands in the snow on camping. Another dog dropped today.
The cloud effects over the land have been very fine all day, rolling cumulus clouds among the snow mountains and deep shadows, alternating with bright sunlight. It was a great relief to us today to plod along with no worry from the dogs. They all followed at their own pace. We had spells of conversation and long spells of silence, during which my thoughts wandered on ahead of me to the days that are yet to come. Our position today at noon was 81°20′S. about 50 miles from Depot B.

Thurs 8 Jan   Turned out at 6 a.m. It had been snowing all night, wet soft flakes at such a high temp. that everything was soaked through. When we started off it was blowing very fresh from the northwest, bringing up heavy banks of fog. The surface became very heavy indeed, so that we covered only one mile in an hour and a half. Meanwhile Shackle got snow blind and was in great pain, so we camped, had our lunch and let him go to sleep.

Sledging notes — in full swing: man-hauling

I spent the afternoon sketching, as it cleared to the most perfect weather, hot sun, no wind, clear blue sky and the whole coast quite clear in the sunlight. Very trying to sketch from, even in snow goggles. Shackle slept all the afternoon and woke practically well. The Captain took a round of angles and slept also. We had supper about 6 p.m. and then turned in. We are now about 40 miles from Depot B and still about 200 miles from the ship. Killed another dog this evening to feed the others.

Fri 9 Jan A McCormick's Skua visited our camp again today. Turned out at 5 a.m. Clear blue sky and sunshine. Temperature plus 17°F. Shackleton had to go blindfold today. We made about 6 miles before lunch today. Camped at 12.30 when the Captain took a sight putting us at 81°6½ 'S. So we ought to reach Depot B on Tuesday, with four days' food in hand. We sailed today with a good southerly breeze all day, but over the soft powdery drift patches we had some heavy pulling. We then pulled from 1.30 till 5 p.m., but as our sledgemeter today lost an essential part of its machinery, we discarded it. I was not sorry myself, for I know the thing has been misleading us continually of late, getting clogged with soft snow and under-recording our mileage. We covered 5 miles after lunch and earned our meagre supper. We just long for more food, anything so long as it will fill. I trimmed my beard and whiskers today. We now have four dogs left, which follow as they like. No day for sketching, as the wind chills one through and through, wet as one is all the time from continual sweating on the march.

Sat 10 Jan   Turned out at 6 a.m. Made a good march before camping for lunch, but unfortunately the sun disappeared before the Captain took the noon sight, and the land which I had been hoping to sketch also disappeared under low clouds. We were then very rapidly buried in a southerly blizzard, whose only recommendation was its direction. So we at once set sail and all we had to do was to hang on to the sledges and prevent the blizzard from running away with them. It was very fast work and the discomfort was great, because the snow and drift were blinding and we soon got soaked as the snow fell practically at melting point. We ran before the gale till 6 p.m. and covered a good deal of ground.
We were wet through when we turned into our sleeping bags, and I steamed in mine like an engine. Warm and comfortable, but so damp! In the morning there were pools of water on the floor cloth where we had been lying and our sleeping bags were wet inside and out. We must have covered 12 miles today. I dropped asleep straight off. There was a shocking drip all night through the roof of our tent, this stuff being good for wind and snow-drift, but no good for water. I killed another dog today, as he had got too weak to walk.

Sun 11 Jan   First Sunday after Epiphany. Turned out at 6 a.m. Still blowing hard, snowing hard and drifting hard. But all from the S.S.W., so we put our backs to it after breakfast and made use of it over a very bad surface for nearly eight hours, stopping an hour at noon for lunch and a sight which put us at 80°44'S. about 10 miles only from Depot B. We have been covering the ground faster than we thought. The discomfort today was intense. We were wet through from melted snow and everything on the sledge was wet through. The drift hid everything from our sight. No sign of land or coast or sky, all drift. A sick looking sun appearing fitfully sufficed to give us our direction, with the wind and an occasional look at the compass.

It was not a pleasant day's work, and towards the end of the afternoon the wind dropped so low and the surface got so very heavy, that we were making less than a mile an hour for all we could pull. We camped at 5 p.m. had our supper and then the sun came out and things began to clear all round, not low enough however to show the land. Clothes very damp. My jacket stiff with ice in parts as I get into my sleeping bag to thaw out, trousers wet through, but there! in less than a month we may be reading our home mails. One more dog dropped today.

Mon 12 Jan   Turned out at 6 a.m. in full hopes of reaching Depot B before night. It was clearing all round. Sun out, no more snow falling. Surface very fair and a fair southerly breeze all the morning. At noon the Captain took a sight and made us in 80°37′S. So we must have made good 7 miles since yesterday at noon.

The bearings on shore are gradually working into position on our left but the moment we started in the afternoon up comes a mild snow blizzard again from the south and down comes its obliterating pall over everything, even the sun. So we had to shove along by the wind's direction only, corrected by an occasional halt to look at the compass. From 1.30 to 3.30 we plodded along, seeing nothing and making a mile an hour, so heavy was the going. Then we camped and decided to wait until it cleared. We have now only two dogs left of our original team of nineteen, and these two walk loose doing no work. We are now in our bags writing up diaries, etc. If only the weather would give us a chance, we could now find our depot in a few hours, but in such thick weather as this we couldn't hope to run across it, and we could easily run in amongst those crevassed pressure lines that we know are close by us now. Our tent drips very badly inside, when this soft snow is falling at high temperatures and we are in a chronic state of dampness. Shackleton has been getting very short of breath for a day or two[11].

Tues 13 Jan   We were up many times during the night to see if it was clearing, but to no purpose. At 5 a.m. we turned out into thick blank whiteness. No land, nothing visible anywhere. We had a reduced breakfast, because we have only a few days' food with us now, and it has to last us till we can find this depot, which may be some time if this weather lasts. Starting at 8 a.m. we made some 4 miles by 11 a.m. with a slight fair breeze, but heavy pulling. Very depressing work in this blank whiteness.

We then lunched on short rations and turned into our bags, deciding to wait and do another hour or so if it cleared before evening. We are so near the depot that now we are afraid we may pass it. About an hour after lunch the fog began to lift and there, not 2 miles away, was the flag we had stuck up marking our food supply. We were up and off at once and after 2 hours' very heavy pulling we reached it. We have with us now two sledges and they have about even weights on them, so we tried them for a hundred yards separately, the one on wooden runners, the other on German silver. And the difference in the ease with which they ran, decided us to take the metal runners off at once and for ever, at summer temperatures at any rate, and on snow surfaces. On blue ice or hard sastrugi, or at the low temperatures of spring and autumn very probably the German silver is the best.

We are now camped at the depot talking over what has to be done tomorrow in the way of dropping all useless weights, re-packing the sledges and what to do with these two useless dogs. Shackleton looks decidedly ill today and is very short of breath. We arrived here with just 407 lbs in all.

Wed 14 Jan   Turned out at 6 a.m. to a fine warm overcast day. No wind. Numberless jobs — digging out our month's provision and cleaning the ice off the bags, recharging our provision tanks, oil cans, rearranging weights and gear. We threw away everything we possibly could dispense with, as we are bound now to reduce our weights, especially as Shackleton seems so seedy. We left clothes, boots, a lamp, straps, two sledges, a large quantity of dog food, kit bags, gloves, socks, burberries, ice axes, alpine rope, and goodness knows what not besides. It took us till 2.30 making up our new loads on two 9 foot sledges. Then we lunched and started off on the wooden runners. We had 288 lbs on the after sledge and 236 lbs on the forward one. We had left all our ski and ski poles and ski boots except one pair or set, for emergencies, in case one of us had to go on ahead and fetch relief from the ship, or to ease anyone who happened to go sick, as it is far less fatigue in this snow travelling on ski than on foot.

We threw off 25 lbs of dog food this evening, and I killed the remaining two dogs, as we didn't see our way to getting them home alive, and they are utterly useless as regards work. So now we have to pull 525 lbs in all.

I had made systematic examinations of all three of us since we started on this journey and now there is no doubt that we all have definite, though slight, symptoms of scurvy. Since the last blizzard Shackleton has been anything but up to the mark, and today he is decidedly worse, very short winded, and coughing constantly, with more serious symptoms which need not be detailed here, but which are of no small consequence a hundred and sixty miles from the ship, and full loads to pull all the way. We camped at 4 p.m. and talked things over in our bags. It is now 8.30 p.m. and I must write up my rather unsatisfactory medical report.

Thurs 15 Jan   Shackleton had a very bad night, but we turned out at 6 a.m. as our object now is to get back as soon as we can. Marched from 8.30 till 12.30. Camped an hour for lunch and then did 3 hours in a sunny fog with nothing in sight but a fogbow. The evening and all night was overcast. No land in sight all day. Very disappointing, as we badly want to take angles, bearings, photos and sketches. Shackleton had a bad day and was not allowed to pull. We all went down a crevasse together today which rather surprised us, but we held up on our elbows and harness. Between us and the shore we can see miles of broken, irregular, hummocked up ice. In the evening I made a sundial and learned something about rope splicing. A skua again visited us. Shackleton had a better night.

Fri 16 Jan Turned out at 6.30 a.m. Overcast and foggy. Not a pleasant day. Cold headwind from the north, but we marched from 9 till 1 over a surface which was everywhere crevassed and broken by large mounds and ridges of pressure. Camped for lunch and marched again from 2 till 5 p.m., then camped for the night. Continually one of us put his foot down a crevasse, but they were all narrow and no harm came of it. We had nothing to look at all day but blank grey cloud and snow surface, no land in sight. Shackle looks very poorly indeed and walks along in his harness, forbidden to work either on the march or in camp. Now that we are covering good distances up to 8 and 9 miles a day, we allow ourselves full rations, which though not filling are enough to work on. When we camped for the night it suddenly began to clear. Land appeared and sunshine.

Sat 17 Jan   Turned out at 6.15 a.m. again to an overcast sky and complete absence of anything to look at, or steer by in sky or Barrier surface. We marched 4 hours in the morning, camped an hour for lunch and then did three hours more. Overcast all day. Very depressing, but we made good progress all the same using a little tuft of wool on a cane as a wind vane to steer by. We saw nothing all day except huge pressure ridges as we came to them and had to haul our sledges over them. Crevasses abundant, but all filled in except at the edges, where there is only breadth enough to drop into and hang up by your elbows and harness. One very rarely puts more than a foot in or a leg, occasionally both. Surface very fair for wood runners.

Shackleton better today, but still very short winded and breathless, not allowed to work either camping or marching. It was fairly close in the morning and we got wet through with perspiration, making the afternoon very uncomfortable, as it became a good deal colder. No sun today, therefore no sight. All our socks etc. which we hang out to dry every night, get smothered in a thick mossy rime of hoar frost in this overcast weather. We made some 9 miles today.

Sun 18 Jan   Second Sunday after Epiphany. Turned out at 6 a.m. to another hopelessly overcast day, with nothing to steer by or fix one's light-dazzled eyes on. After an hour's marching we found the steering so erratic that we camped and waited to see if it would clear. At 4 p.m. there was enough break in the cloud ahead to steer by, so we had another breakfast and at 6 p.m. started off again. We did three hours and then camped as Shackle was feeling bad. Now it is nearly midnight and we have turned in. Outside one can see nothing at all, either in the sky or below it, all one uniform brilliant grey light without a break. One cannot see one's own footsteps in soft snow in this light, nor any of the inequalities that one stumbles over. All our scurvy symptoms are improving on an increased allowance of the dried seal meat and no bacon at all, the item of our diet which I believe is mainly responsible for our scurvy. My eyes today have been very weak and painful.

Mon 19 Jan   Turned out at 8 a.m. Still overcast and dismal. Not a vestige of land in sight. We started off about 11 a.m. and at noon the Captain managed to get a sight which put us in lat. 79°58′S. We trudged on for three hours more and camped for lunch. After this we did another three hours, making about 9 miles in the day again. Very depressing this white pall, and very disappointing to be forced to keep going on and on, knowing that we are missing for good our opportunities of sketching and surveying. Not once have we seen the land clear since we left Depot B. It is now 10.15 p.m. and we are in our bags for the night. Shackle very much better, but allowed to do nothing but just walk along. We cannot carry him. The moment he attempts a job he gets breathless and coughs. The Captain and I can quite well manage everything alone and the surface happily is excellent, so that the sledges run easily.

Every half hour today the characteristic 'hush' of the snowcrust settling over large areas as we passed over it was to be heard. My eyes are still very painful and I walked all day with a worsted balaclava drawn over my face, which allows one to see where one is going. Temperature plus 15°F. and fairly chilly except on the march when one sweats freely.

Tues 20 Jan   Turned out about 8 a.m. and all the morning the sky was clearing with a light breeze, till at midday we could see land appearing here and there, in a very dazzling white light. It was a huge relief to all of us at last seeing sunshine again and feeling its warmth. The Captain took a noon sight which put us at 79°51′S. We camped after 4 hours' marching over rather a heavy surface, had our cold tea and lunch and started again. After an hour the land had all so far cleared that we halted and pitched camp.

It was a very grand sight indeed in the sunshine after our long spell of thick weather. The Captain wanted to take a full round of angles for his survey and I wanted a sketch of the whole show. It was bitterly cold work sitting for an hour and a half in a cold southwesterly breeze with the temperature only half a degree above zero. Shackleton cooked our supper while we were outside and after supper we did another hour's work, and then had some very hot cocoa before turning in, for we were very cold indeed. However we had got what we wanted and we were glad, because no sooner had we turned in than the sky again became overcast, and an hour or two after we had finished our work the whole thing was blotted out again by a southwesterly blizzard. We saw Mount Discovery today and a mirage of the Bluff.

Wed 21 Jan   Blowing a pretty stiff breeze when we turned out at 7 a.m. Sun appearing fitfully between very stormy looking clouds. Had breakfast, struck camp and were away with floor cloth sail set to the fair wind at a great rate. Shackle at first was walking in harness, but we made him sit on the after sledge and break its pace with a ski pole, as the sledges were going too fast for us. This worked for an hour or two, but not well so we made [him] walk behind at his own pace, while the Captain hauled back on the port side of the front sledge and I on the starboard side of the after sledge, and so we guided them and went gaily along. Camped at 1 for lunch hour, chilly and cold, then off again under sail for three hours more, camping for supper about 6 p.m. Mount Discovery and the Bluff are in sight again now, so we feel we are nearing home. The wind held all day, but dropped in the evening. The barometer however is exceedingly low. There has been no snow fall with the wind today and but little drift. We must have covered nearly 12 miles today.

Thurs 22 Jan   Turned out about 7.30 a.m. and found a blizzard blowing, so we covered ourselves up in burberry, shoved all the breakfast gear in the tent and cooked our food. It was drifting freely and very chilly, but as soon as we had finished breakfast the drift stopped. So we packed up and started off before a strong south westerly breeze, which carried both our sledges along by themselves. We only had to guide them. Shackleton followed on ski at his own pace. We ran on so for 4 hours and camped for lunch. Noon sight put us in 79°34′S.

After lunch we did three hours more at a more moderate pace, the wind gradually dropping towards evening. The sky cleared and the barometer began to rise. We are now about half way between Depot A and B. A week more should bring us into A. No land in sight today, so no sketching. We have a desperate hunger. The surface was very heavy today, but the wind made up for it. Now in our bags, 11 p.m. My eyes so weak lately and so ready for sleep that my reading hasn't come off for a good many nights. These are fatiguing days and one sleeps before one has toggled up the bag almost.

Fri 23 Jan   Turned out about 8 a.m. and with a good southerly breeze we started off at a grand pace. Shackleton went on half a mile ahead with a compass and gave us the course, on ski, the Captain and I being left alone to pull or guide the sledges under full sail all day. Forenoon 4 hours, afternoon 3 hours is our routine now, with an hour for lunch, when we pitch the tent and sit on our bags rolled up. Chilly and grey and overcast. Nothing to fix your eyes on in earth or sky, except a few indistinct rock patches along the coast on our quarter, half hidden in mist. Very disappointing all this thick weather. The Captain and I had long talks on every subject imaginable and indeed he is a most interesting talker when he starts. Weather just as thick this evening. My eyes very weak again today from the grey but brilliant light, a light that strains one's sight terribly.

Sat 24 Jan   Turned out at 7 a.m. Overcast still, but a fair southerly breeze blowing. Started off about 10 a.m. and did a good march by 2 when we camped and had lunch. Still overcast, but we went on till between 6 and 7 p.m., making some 10 miles good towards Depot A. Shackle went ahead all day on ski, giving us the course by compass, as there was no sun and no land in sight and absolutely nothing to steer by. Occasionally we got a glimpse of the Bluff. Early in the afternoon we had three skuas round us. The Captain and I, as we plod along with the sledges, had long and

Three men inside a pyramid tent

interesting talks. The snow is very deep and the walking very heavy today, so that we regret that we threw away our ski. However as they weighed some 25 lbs perhaps it was as well. My right eye pretty bad this evening.

Sun 25 Jan   Third Sunday after Epiphany. Turned out about 7 a.m. to a very glorious sunny Sunday morning, the whole coastline in sight now in brilliant sunshine. We were in high glee in consequence, as it gave us just the opportunity we had been longing for, of joining up this coast with what we got on paper some days ago. The Captain

71

took a round of angles and we then did 2 hours on a heavy surface, into which we sank deep at every step, but over which the sledges ran easily. Slight northerly head breeze. We then halted and the Captain took a noon sight, putting us in lat. 79°13′S. We are therefore about 35 miles now from Depot A and about 95 miles from the ship. We then did 1½ hours more and camped for lunch, when we had hot tea for a luxury and I was given three hours to sketch, while the Captain took a round of altitudes and bearings. We then marched 2½ hours more, making 6 in all and camped for the night. We saw our old friend Mount Erebus again today for the first time, his smoking head just appearing about the Bluff cape. In ten days or a fortnight now we may be back at the ship. We are having fuller meals now, as we have time well in hand, and we no longer dream of food. Indeed we sleep better than we have the whole journey.

My eyes have been in a sorry state all day from sketching with sunglare, streaming with water and very painful from time to time. Sketching in the Antarctic is not all joy, for apart from the fact that your fingers are all thumbs, and are soon so cold that you don't know what or where they are, till they warm up again in the tent (*then* you know all about it!); apart from this you get colder and colder all over, and you have to sketch when your eyes stop running, one eye at a time, through a narrow slit in snow goggles. No one knows till they have tried it how jolly comfortable it all is.

Mon 26 Jan   Turned out at 6 a.m. after a splendid night of unbroken sleep, now the rule with us. The surface we are on is the worst we have had the whole journey, very soft and heavy walking and very treacherous. Started off with a good southwesterly breeze at 9 a.m. and after 4 hours we camped for lunch and a sight. The wind then dropped and in the afternoon we did 3 hours, at a much slower pace. Shackleton went on ski. Though he is very anxious to work again, I don't think it is safe and have forbidden it, though he is certainly a good deal better, but the least exertion makes him breathless. At 5 p.m. we ran across the old tracks of Michael Barne's party making their way to the same depot that we are making for. From the direction, we think he must have made a good journey to the W.S.W. and from the tracks we could see that none of his men were knocked up. The weather today has been glorious blue cloudless sky and hot sun, making us sweat terribly over the heavy surface. My left knee has felt strained at the back since Wednesday last.

Tues 27 Jan   Turned out at 6 a.m. Glorious clear sunshine and blue sky again, the whole length of coast line looking absolutely fairylike in white and pale blue transparency. Got off at 9 a.m. Captain took a sight at noon, 78°57′S.lat. We did another hour and camped for lunch. Although the sun was so hot that it scorched one's skin and one had to wear the large sun hats, the cold in the shade under them was such that our moustaches and whiskers all became frosted and covered with ice, making one's lips and nostrils sting and tingle very painfully, the temperature of the air by sling thermometer being one degree below zero F. There was a good hard crust on the snow, making the walking much easier till the afternoon when the sun had had time to soften it again. We had three hours of heavy pulling after lunch, but on the whole are making good way. Our lips are in a horrid state. The Captain's and mine have been raw for many weeks now. Turned in now at 9 p.m. Still clear and sunny, with a little cirrus flying about in a blue sky.

Wed 28 Jan   Turned out at 5 a.m. Had a regal breakfast with the prospect of reaching the depot before night time. Imagine our disappointment when we came out again after breakfast, not an hour later, at finding the land all blotted out of sight by the drift from a southwesterly breeze and a low stratus that had come up with it. However we set sail and started off and in an hour or two the wind went round to the west a bit and the land began to clear again, enough at any rate to steer by. We did 5 solid hours at a quick pace in the forenoon. Mounts Erebus, Terror, the White Island and the Bluff all in sight when we camped for a luxurious lunch of hot tea. Then off again and in about 2 hours we saw the black flag that marked our Depot A. Cold wind all day, but plenty of sunshine too. Blue sky and low drift, a bad day for the complexion.
Here at the depot were the remains of many camps and here too were letters from the ship for us all. I had long letters from Charlie Royds. Excellent news of all the other sledge journeys, no mishaps. But of course no news of the relief ship, as these were left for us at the beginning of January by Barne's surveying party, which we found were now only 4 days ahead of us on their way to the ship. We are now 60 miles only from the ship and we pick up food here to carry us in in absolute luxury — sardines, port wine, raisins, prunes, chocolate, and all the other substantials such as pemmican, sugar, tea, cocoa, and what not. Sunshine now and the wind dropping, and we are in our bags after supper feeling terribly full.

Thurs 29 Jan   Jim's birthday. We woke up to a real howling blizzard, the air chock a block with snow drift. Shackleton again utterly knocked up with cough and breathlessness, quite unfit to

Young penguins feeding

move out of the tent, so we decided to remain where we were till the blizzard dropped. We lay in our bags, had full meals all day long, read a chapter of Darwin and slept. Shackleton very poorly indeed all day, very breathless, very restless, and quite unfit to move. In the evening the wind dropped and the sun came out, and we found everything buried in snow drifts and many things lost that we should have considered serious a month ago, but now we are only six days from the ship so 'what's the odds'?

Fri 30 Jan   Turned out about 6 a.m. and started digging things out, and packing up for our last stage of the journey. We had a fair southwesterly breeze today and a good surface and we covered good ground for 4 hours before camping for a hot lunch. Shackleton, still very seedy, went his own pace on ski. We then did 5 hours more and camped at 8 p.m. having done some 15 miles in the day. Our main object now is to get Shackleton back to the ship before we get caught in another blizzard. He has been very weak and breathless all day, but has stuck to it well and kept up with us on ski. The Black and Brown Islands have now come in sight.

Sat 31 Jan   Warm and sunny, blue sky, dead calm. Turned out at 7 a.m. Shackleton had a good night and is much better for it. We did 5 hours solid hauling in the forenoon and 4 hours in the afternoon, when we had a very elaborate supper and turned in for the night. My strained knee had an uneasy day of it, as we were on crevassed ice nearly the whole time and I went down about eight of them. We took turns, the Captain and I. Shackleton avoided them as he was by himself on ski. In the evening I did a bit of sketching, one of the finest pieces of mountain scenery imaginable. Lamp soot and the greasiness unavoidable with tent cooking has made us funny to look at.

Sun 1 Feb   Turned out again to a glorious morning of still calm, clear, weather, hot sun and very cold air. At 10.30 a.m. we started off and made 4 good hours steady but heavy hauling all along the coast of White Island. Our camp last night was close to one of the immense Barrier ridges which have to do with the tide crack. Shackle was on ski all day. After lunch we made 3 hours again and camped for the night. Castle Rock and the first ridge now in sight and not much more than 20 miles between us and the ship.

Mon 2 Feb   Again blessed with fine weather. No wind and a very fair surface, broken only by a series of crevasses and cracks as we were rounding the pressure end of the White Island all day. I went down several cracks pretty heavily, but being tied to the sledges by a very strong harness one has no fear of disappearing below. My knee troubles me very little on the march, except at these crevasses where I am afraid of straining it really badly. But in camp it gets very stiff. We camped an hour for lunch, and then after covering some ten miles and getting well on to the north side of the island, we pitched our last night camp and hoped for one more fine day to take us in to the ship tomorrow.
Mounts Erebus and Terror have been a magnificent sight all day on account of some alto-stratus which threw up the sunshine and shadow in high relief over all the snow. They are the most completely snow-clad mountains anywhere near us, hardly a piece of bare rock showing on the southern side, and being 12,000 ft high, their beauty can be better imagined than described. As we near the coastline we get far more greys, with rose pink and light red lights in them in the sky.

Tues 3 Feb   Made an early start, the Captain and I pulling, Shackleton going on ski. The day began rather overcast, dead calm but very close and warm. Moderately good surface, and all the home landmarks well in sight, though Observation Hill and Cape Armitage of course cover the ship from our view. After marching 2 or 3 hours however, we saw ahead of us what we thought was a seal at the edge of the old Barrier ice. It turned out to be the remains of our last year's depot, and before we reached it we had the greater pleasure of seeing two figures hurrying towards us on ski. Just 6 miles from the ship we met them — Skelton and Bernacchi, clean tidy looking people they were. And imagine our joy at hearing that the relief ship *Morning* had arrived a week or more before and that all our mails and parcels were waiting for us in our cabins. All the news was good about everything, except that there were still eight miles of ice floe to go out before we should be free to leave our winter quarters. However that didn't trouble us much.

We camped and had a good lunch and then these two pulled our sledges in for us. Our flags of course were flying and we had a very gay march in, listening to scraps of the world's news, and scraps of our own little world's news, the news of the ship. We had been doubly cut off for three months from any news but what we had brought ourselves from the unknown south. Three miles from the ship we were met also by Sub. Lieut. Mulock, one of the *Morning's* officers, a very nice young fellow who is to join up with us on the *Discovery*. He is an R.N. Officer of the Survey Department. Next we were met by Koettlitz, Royds, and all the rest, and a crowd of men. It was a great home coming, and as we turned Cape Armitage we saw the ship decorated from top to toe with flags and all the ship's company up the rigging round the gangway ready to cheer us, which they did most lustily as we came on board.

They were all most enthusiastic and everyone shook us by the hand all round, it was a most delightful welcome. Charles Royds was quite charming. A lot of photographs were taken and indeed we must have been worth photographing. I began to realize then *how* filthy we were — long sooty hair, black greasy clothes, faces and noses all peeling and sore, lips all raw, everything either sunburnt or bleached, even our sledges and the harness — things one didn't realize before, and our faces the colour of brown boots, except where the lamp soot made them black.

Then came the time for a bath, and clothes came off that had been on since November the second of the year before, and then a huge dinner. Captain Colbeck, Engineer Morrison, Lieuts. Doorly and Mulock were all there, and a long and tiring evening followed. But instead of drink and noise and songs and strangers, I know I was longing to lie down on my bunk and have a long quiet yarn with Charles Royds. I was in no hurry at all to spring at my letters, for I felt an absolute confidence that everything was well with all that I cared for most at home, and the only thing that was real sad news was the dear old Aunt's paralysis. Such was our home coming after an absence of over thirteen weeks.

The "Discovery" in Winterquarters. 1903.

EPILOGUE

Thus ended the first long-distance journey to be carried out in Antarctica — 960 miles covered in 93 days. Scott's high hopes of exploring the mainland and finding a direct approach to the Pole had not been achieved; dog transport had turned out to be a failure; and Shackleton's health had been seriously undermined. Nevertheless, for Wilson the trip had not been altogether without its rewards, as his numerous sketches, including over 200 feet of panoramas of newly-discovered coastline, bear witness.

Towards the end of February 1903 it was found that the *Discovery* was immovably frozen into her winterquarters and the relief ship *Morning*, bearing with her the invalid Shackleton, had to sail northwards unaccompanied. Wilson soon recovered from the effects of the sledge journey and settled down happily enough to take advantage of a second Antarctic season. There was so much to be done including the sketching of scenery, wildlife and the fascinating, everchanging, elusive aurora. Above all, there was an opportunity for Wilson to return to his 'proper sphere of work' — namely the breeding habits of the Emperor Penguin and the importance of its embryology to evolutionary science. Two visits to the rookeries at Cape Crozier in September and October proved only partly successful, and, at temperatures of around −62°F., distinctly uncomfortable. 'It has been worth doing,' he wrote, 'but I am not sure I could stand it all over again.' Yearning to be reunited with his beloved wife, Wilson could hardly have foreseen then that only a few years later he would deliberately repeat the journey, only this time in the middle of winter and in considerably more uncomfortable circumstances.

Emperor Penguins at Cape Crozier

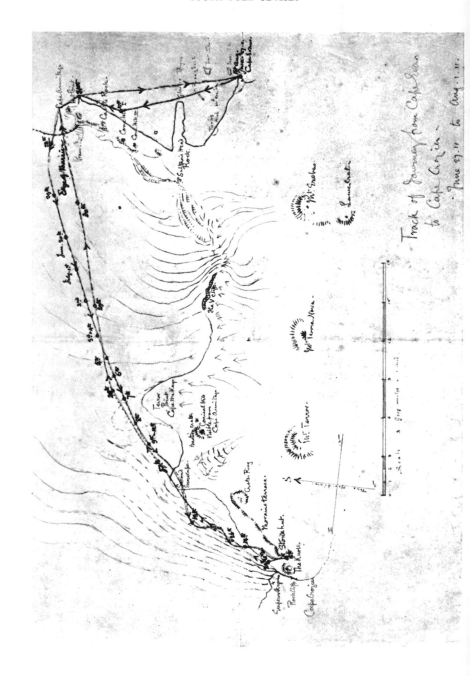

Edward Adrian Wilson by *A.U. Soord*

Sledge hauling on ski

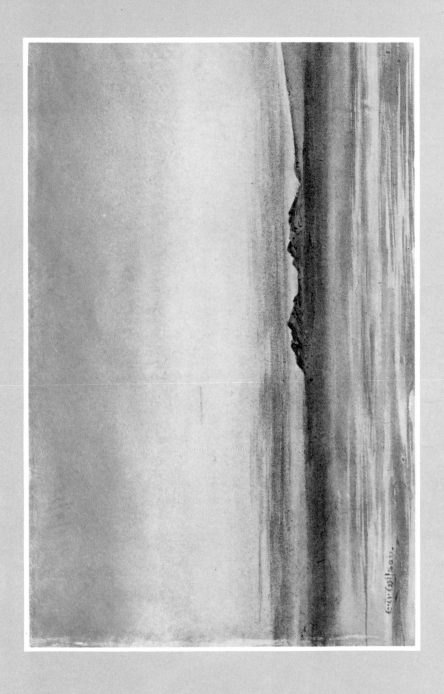

Furthest south. December 1902

The *Discovery* in Winterquarters

Emperor Penguins

Exercising the ponies

Mount Discovery and Inaccessible Island

Paraselena. Jan. 15. 11. 9·30 pm.
Cape Evans. McMurdo Sound.

Paraselena. Cape Evans. McMurdo Sound

3

THE WORST JOURNEY
IN THE WORLD

PROLOGUE

I T was on board the *Terra Nova* in December 1910, and a day or two before entering the pack ice which guards the approaches to the Ross Sea, that Captain Scott and Wilson met to discuss the problem of where to set up their winterquarters. There seemed to be two possibilities; Cape Crozier on the eastern extremity of Ross Island, and Cape Royds, a promontory forming the western extremity of Ross island. Cape Royds had been used successfully as a base by Shackleton during his *Nimrod* expedition of 1907-1909, but Cape Crozier had the twofold advantage of easy access not only to the Barrier, but also to vast rookeries of Adélie and Emperor penguins. And on this occasion Wilson was determined not to be cheated of his ambition to obtain specimens of the freshly incubated Emperor Penguin eggs. It was his firm belief that a study of the embryonic development of this most primitive bird might throw light on certain evolutionary problems, such as whether feathers have evolved from reptilian scales, or whether they have an independent origin. The specimens obtained during the *Discovery* expedition had proved useless for analysis. Fresh eggs laid in the depth of the Antarctic winter were essential for Wilson's requirements, and Cape Crozier was adjacent to a major rookery. These plans were soon to be dashed, for, when in early January 1911 the *Terra Nova* at last arrived off Cape Crozier, it was found that the strong swell made any sort of landing impossible. Cape Royds, cut off from the south by open water, proved equally unsuitable, and so, at last, a cape of moraine and rock, known from *Discovery* days as 'The Skuary', was chosen instead. Renamed Cape Evans, after Lieutenant Edward Evans, Scott's second-in-command, it was by no means ideal. To reach the Barrier and the route to the Pole, a journey over the sea ice of McMurdo Sound was involved, and it was a good sixty miles distant from the Cape Crozier penguin rookeries. If Wilson's plans were to be realized, a hazardous journey in darkness and at temperatures which might be expected to touch −70°F. would be necessary. Captain Scott might well have been forgiven for forbidding so hazardous an undertaking. A sledge journey in the Antarctic midwinter had never been undertaken before, and to attempt one at a time when it was essential to conserve his best men and resources for the forthcoming Pole journey must have seemed to be tempting fate. As it was, the expedition had been dogged by a succession of minor disasters and here was Wilson planning to put at risk not only himself but two of Scott's best sledgers, the young scientific assistant Apsley Cherry-Garrard ('Cherry'), and Lieutenant Henry Bowers ('Birdie') of the Royal Indian Marine, described by Scott in his journal as 'the hardiest traveller that ever undertook a polar journey'. But this birdsnesting expedition was as close to Wilson's heart as the Pole was to Scott's, and it is almost certain that Wilson had made its realization a precondition of his accompanying Scott on this second trip south. Scott having finally impressed on Wilson that he 'had the pick of the sledging element and mustn't get them crocked' finally gave his assent.

Shortly after a boisterous Midwinter Day celebration Wilson, Bowers and Cherry-Garrard made ready to depart, and Scott noted in his diary: 'This winter travel is a new and bold venture, but the right men have gone to attempt it. All good luck go with them.'

DIARY

Tues 27 June   Leaving the Hut at Cape Evans a little before 11 in the morning after being photoed with our sledge in the dark by flashlight, Bowers, Cherry-Garrard and I started off for our first day's march, accompanied by Simpson, Meares, Griff Taylor, Nelson and Gran, who all helped us to drag our two sledges. A number of others came to see us off round the Cape. Nelson and Taylor left us when we had gone 3½ miles. We continued with the other three. We made in to pass as close as possible to the end of Glacier Tongue where there were said to be fewer pressure ridges in the sea ice. It was so dark, however, that we never saw Glacier Tongue and we only knew we had passed it when we saw Turk's Head disappear. We then ran in to some very bad hummocky ice, and our rear sledge capsized. It was too dark to avoid them, so Meares, Simpson and Gran helped us on till we were again on smooth ice. At 5¼ miles they also left us, and returned to Cape Evans. The loads were heavy enough now, on sea ice even, to make us slow, though the surface was good. We camped for lunch at 2.30 at 6⅓ miles from Cape Evans. We were using a lined tent which was an invention of Sverdrup's[1] in his last expedition, and we found it a great boon for it undoubtedly made one more comfortable in camp at the low temperatures we were to have this journey. It wants care and we made a point of brushing all the hoar frost off it every time we struck camp. The steam of the cooker otherwise gradually collects in the tent and ices the whole inside up. It was the duty of the cook for each day to see to this, and we were cooks each for one day at a time in turn. The lowest third of our tent, as a matter of fact, became badly iced up, but the upper parts we managed to keep clear of ice.

Getting away at 4 we made for what we believed to be Hut Point, but in the dark got a good deal too close in to Castle Rock. Our pace was slow owing to the weights and we got into an E.S.E. wind which blew force 5 till we camped at 8 p.m. Then the wind dropped and we had a clear starlight night. We had purposely started several days before the moon would rise in order to have all our moonlight later when we got on to less well known country than this. The temperature for the day ranged from −14.5°F. to −15°F. and the minimum for the night was −26°F. So ended our first day out, and perhaps I should say here why we had started on a sledge journey of such length. We hoped to be away 6 weeks in the darkest part of the whole year, instead of remaining comfortably in the hut, simply because the Emperor Penguins at Cape Crozier laid their eggs, as far as we could judge from what we found out in the *Discovery* days, in June and it was up to me to go and collect some of them to get early embryos − as I have said − for microscopic work. If vestiges of teeth are ever to be found in birds of the present day it will be in the embryos of penguins which are the most primitive birds living now, and the Emperor is quite the most interesting of them all, and the most difficult to get at, as will appear shortly from my account. Travelling with sledges in this way in mid-winter had not been done before, I believe, so it was an experiment of some interest. I had the pick of the whole party, Birdie Bowers and Cherry-Garrard as my companions whom I had chosen and who were allowed to come by the Owner[2] on condition I brought them back undamaged. They were the best of all our new sledging lot.

Wed 28 June   Turned out 7.30 a.m. The going became heavy and we made little more than a mile an hour. Surface rough sea ice. We reached Hut Point at 1.30 and lunched there. After lunch made better going to Cape Armitage and then had the only really good 2 miles' going that we met with the whole of this journey, and soon reached the edge of the Barrier finding a good slope up and having no difficulty in hauling the sledges up one at a time. There was a snow covered crack at the top of the drift invisible until stepped into, but we knew it would be there and so had no difficulty. Coming down the slope of the Barrier was a steady stream of very cold air which we noticed only a few yards from the bottom and lost a few yards from the top. It was now 6.30 p.m. and we camped at 7, the last half hour being uphill on the Barrier and very heavy dragging compared with the sea ice we had just left. Temperature ranged from −24.5°F. in the morning to −26.5°F. at Hut Point, and −47°F. at the edge of the Barrier.

Thurs 29 June   A cold night with temp. down to −56.5°F. and it was −49°F. when we turned out at 9 a.m. in pitch dark with a candle in the tent. But the day was fine and calm with occasional light easterly airs and aurora in curtains to the east both morning and evening covered the greater part of the sky. One of our chief pleasures on the march eastward was to watch these changing auroras for hour after hour ahead of us − almost always in the east and south east, and up to the zenith over our heads. The temperature remained at −50°F. all day, and we felt the cold a good deal in our feet on the march, Cherry getting his big toes frost bitten and I my heel and the sole of one foot. A good many of Cherry's finger tips also went last night and are blistered this morning, but he takes them all as a matter of course and says nothing about them. The surface all day very heavy and made our progress very slow.

Fri 30 June   The surface today was too heavy for us with both our sledges, so we relayed from 11 a.m. to 3 p.m. by what little daylight there was, and from 4.30 to 7.45 by candle lamp. We thus took one sledge on at a time and came back for the other one. The surface was just like sand at these low temperatures. We made only 3¼ miles today in all, but walked about 10. The temp. ranged from −55° in the morning to −61.6° at lunch, and −66° on camping at night. Calm weather and aurora.

Sat 1 July   Turned out 7.30 a.m. From 10.45 till 3 we were able to do relay work by the daylight. After lunch we continued with a candle lamp till 7.45 p.m. Surface like sand and the dragging so heavy that we could only just manage to pull one sledge by itself now. Subsidences in the crust were frequent all day. We made in all 2¼ miles today − very slow work. There was a fine aurora 5 to 7 p.m.

Minimum temp. last night was −69° and today it ranged from −66.5° to −60.5° at 10 p.m.

Sun 2 July   Third Sunday after Trinity. Minimum for the night −65.2° with a breeze of force 3 from S.S.E. with slight drift. Temperature during the day ran from −60° to −65° with calm and light airs which forced us to use our nose nips, for the slightest breeze with such low temperatures freezes any exposed part of one's face at once. We turned out at 7.30 a.m., relayed 11 to 3, then lunch and again relayed from 4.30 to 8 p.m., this time by moonlight, the rising moon giving us just enough light to go by. At one time it passed exactly behind the crater of Mount Erebus and looked like a magnificent eruption. We made only 2½ miles in the day — very heavy sandy snow.

Mon 3 July   Minimum temperature for the night −65°. Sky becoming overcast with some storm clouds on Mount Terror — 1½ miles by day glimmer and by moonlight after lunch. Temp. during the day ranged from −52° to −58.2°. We had a magnificent aurora in the evening when long swaying curtains almost covered the sky up to the zenith where they all became foreshortened, as though hanging right overhead, and swinging round from left to right in a rapidly moving whirl, constantly waxing and waning in brightness in different places, with bands of pale orange and emerald green and lemon yellow colour rippling along the borders of the curtains, and all fading upwards into the darkness of the sky as though they were curtains hanging from the roof of a vast dark cave stirred by some winds which didn't reach us on the surface. It was so very striking that we all three lay on our backs in the snow and watched it till we were too cold to watch it any longer, but we had a great part of it ahead of us as we marched. Our sleeping bags are beginning to get wet thanks to these low temperatures. One has to shut oneself right in all night to sleep at all and one's breath of course gradually makes the bag wetter and wetter. So all the sweat of seven or eight hours heavy hauling each day freezes in one's clothes and this also thaws into the bag every night. Cherry's bag is a large one. My own was a good fit on the small side — it became too small when it was really wet and frozen later on, and it broke at each end and nearly in two across the middle and gave me some beastly cold nights. Birdie's bag was the right size for him and lasted well. We also took with us eider down, or rather ducks' down, sleeping bag linings as a standby, in case our reindeer sleeping bags became impossibly wet and frozen. Cherry was so cold in his large bag last night that he began his down lining today. Birdie and I were as warm as we expected to be in our reindeer-skin bags. He has his with the hair outside. I have mine with the hair inside.

Tues 4 July   Min. temp. for the night −65.4°, but we turned out at 7.20 to find the sky all over-cast and snowing with a gusty southeasterly wind. At 9.30 a.m. the temp. had risen to −27.5° with a wind of force 4 from the N.E. Nothing could be seen by which to steer a course so we had breakfast and turned into our bags again to wait. We were fairly warm and comfortable all day and it began to clear as night came on.

The min. temp. for the day was −44.5° and in the night it again fell to −54.6°. Clouds entirely obscure Erebus and Terror. This day in our bags has saturated our clothing and our bags so the moment we leave the tent we are frozen stiff into a sort of tin mail. It is difficult to realise how clumsy all one's movements become when all one's clothing is frozen stiff right down to the vest and shirt nearest the skin. They are the only articles which one cannot shake a cloud of hoar frost out of when one returns to the hut after a cold journey at these temperatures. The clothing I was wearing on this journey was as follows: on my head a balaclava of close fitting woollen stuff with a windproof covering. A muffler − and these covered one's head and neck, only the face showing, nose, eyes, mouth and cheeks which could be further covered by a face guard buttoning across − but I only had to use mine twice the whole 5 weeks we were out. On my body I had a very thick woollen vest, a thin woollen shirt, a knitted cummerband reaching from the chest to the hips with canvas shoulder straps to keep it from slipping down. This is a splendid article − Ory made me several of them. Over this a grey woollen sweater and over all a wind-proof gaberdine blouse. On my legs a pair of thick pants − thick woollen pyjama trowsers − and over all windproof gaberdine trowsers. On my feet 3 or 4 pairs of socks and reindeer skin fur finnesko. On my hands half mits, leaving the fingers and thumb bare − these never came off − with short removable mits to cover the thumb and fingers, and a large pair of fur mits hung round the neck with lamp wick. With all this clothing wet and all frozen stiff it becomes difficult to move with one's customary agility and any climbing up ropes or out of crevasses becomes exceedingly difficult for anyone but an acrobat. Our feet have so far been warm except in the heavy soft surface snow on which we make such slow progress on the march. One's finnesko get frozen into stiff leather boots and lose all their value in giving the feet free play. So I had to keep a very watchful eye on 3 pairs of feet and continually when one asked if they were cold, it was only to be told they had been cold, but the owner didn't think they were frozen as they were most comfortable. I had to judge then whether they were really warmer or actually frozen, and sometimes it was exceedingly difficult to tell what was happening even with one's own feet. Happily we had no more than superficial frost bites on our toes and heels and soles, but every now and then we had to camp and have our lunch an hour earlier on account of this. We couldn't afford to risk getting anyone crippled in the feet above all else.

Wed 5 July   At 7 a.m. we turned out and the surface was the worst for pulling on that we had yet had. We relayed for 8 hours and only advanced 1½ miles in the day. The min. temp. last night −54.8° and by the evening it fell to −60.1°. We are surrounded now by white fog, but we can see Erebus and Terror and the moon with a halo and mock moons and vertical shafts of the lunar cross. We were hauling up hill all day crossing Terror Point, a long snow cape.

Thurs 6 July   Calm and clear, though a heavy fog bank lies over the pressure ridges on our right and over the seaward area ahead of us. Relay work again for 7½ hours in which we made 1½ miles advance. The min. temp. for the night was −75.3°. At starting in the morning it was −70.2°. At noon it was −76.8°, and at 5.15 p.m. when we camped for lunch it was −77°. At midnight it rose again to −69° with some low lying mist. These are very low temperatures, about 10 degrees lower than any we got in *Discovery* sledging. This was the lowest we got −77° i.e. 109 degrees below freezing point Fahrenheit. Amundsen had lower temperatures in the north but it wasn't sledging as we are. I shall be very much surprised if he isn't beating all temperature records where he is now on the edge of the Barrier at the Bay of Whales.

Fri 7 July   We got away late at noon in a thick white fog which made it impossible to see where we were going. We still had to relay though the surface was certainly improving. We had 4 hours between noon and 4, then camped for lunch and then went on till 9.45 p.m. when the fog was so thick that we couldn't see where we were going. We made only 1⅔ miles in the whole day. The min. temp. last night was −75.8°. At 2 p.m. −58.3° and at 7 p.m. −55.4° a rise for which we were grateful as it is distinctly easier to save one's feet from frost bite at −50 than at −70.

Sat 8 July   A day of white fog and moonlight but no sign of land to steer a course by. We relayed again, 4 hours for 1¼ miles in the forenoon and 3 hours for 1 mile again after lunch. Surface improving as we get into a more windy area. Min. temp. for the night −59.8° and by the evening it had risen to −47°.

Sun 9 July   4th Sunday after Trinity. Relaying impossible owing to the fog and falling snow which covered our footprints and made it impossible to find the sledge load left behind. We found that we could drag the two sledges together, however, today. We couldn't see where we were going as everything was obscured by fog. Our chief difficulty now was to avoid running unwittingly up on to the Terror slopes on the left where there are many patches of crevassed land ice, and into the Crozier pressure ridges on the right. No landmarks visible. We advanced about a mile in the forenoon march and ¾ mile after lunch — the moon invisible. In the after lunch march we ran into crevassed ground after having suspected we were pulling the sledges up and down over several rises.

Pulling in the dark it is almost impossible to know whether the pulling is heavy because one is on a bad flat surface, or because one is pulling up hill. We expected some rises before reaching the second long snow cape beyond Terror Point, so we went on. The surface was again hard and icy in places with sometimes 6 inches of snow on it, as it was also yesterday. Our feet often went through the soft snow and slipped on the hard wind-swept sastrugi underneath. This was on the top of the ridges. In the hollows the snow was deep and soft. One could judge a good deal of the nature of the surface and the chance of running into cracks by the sound and by the pitch of the note which one's feet made on the more exposed snow — a much higher pitch than the lower note of deep and safer snow — and this in the dark without seeing what one was walking on at all.

Occasionally the moonlit fog caught the edge of a snow ridge, but otherwise one saw nothing. We appear to be still out of the run of the main blizzard winds for which Cape Crozier is famous, and in an area of eddying winds, snowfall and fog. We call it Fog Bay. The moon became visible overhead but nothing else until just as we found ourselves going up a longer and steeper rise than usual we saw an irregular grey mountainous looking horizon ahead of us loom through the fog. It was evidently close up. So we unhitched from the sledges and tying ourselves together by our sledge harness traces we walked up a long hard snow slope to the top of our mountainous horizon and then found we were on one of the large Crozier pressure ridges with another appearing over the fog again. This was more certain to us when we stood still and heard the ice creaking and groaning ever so deep down under us and all around. This excursion from the sledges gave us our proper direction for safety, as we thought, but after retracing our steps and then turning more to the left, i.e. north, with our sledges, we again ran in amongst pretty big pressure ridges with cracks and crevasses in them, and so finding a hollow with deeper snow in it we camped for the night and decided to wait until we could see exactly where we had got to. The question of tide crack was in our minds all the time while we were crossing the long rounded snow capes which run out from Mount Terror into the Barrier, notably at Terror Point or at Cape Mackay, and the 'Second Cape' further on towards Cape Crozier. We certainly did cross several small cracks on these slopes which had the appearance of a certain amount of working, but they were inches only and if they are tidal they must take up very little of the tidal movement. Probably nine-tenths of the tidal action is taken up by the large pressure ridges. Today the temp. ranged from $-36.7°$ up to $-27°$, with light airs. Some hours after midnight it began to blow and snow began to fall heavily while the min. temp. for the night was $-24.5°$ up to noon of Monday,

Mon 10 July   when a blizzard was blowing from S.S.W. force 6 to 8, and the air was as thick with snow as could be. This continued all day and we lay warm and wet in our bags listening to the periodic movements of the ice pressure underneath our tent and all around us.

Tues 11 July   The temp. at 10 a.m. went up to $+7.8°$, and at 8 p.m. was still $+6.8°$. The lowest all day was $+3.2°$. The wind was southwesterly force 5 to 9, very squally, and this continued all day with a heavy snowfall which packed our tent about 1½ to 2 feet all round, as well as burying our sledges. Cherry sleeps in his down bag lining inside his fur bag. I turned my bag from fur inside to fur outside, the high temperature and the long lie in during this blizzard has steamed us and our bags into a very sodden wet condition. One wonders what they will be like when the temp. goes down again. We have discussed our respective rations and they have been revised somewhat as follows. To begin with we all three decided on simplicity, so we cut our foods down to biscuit, pemmican, butter, and tea — nothing else. We cut out sugar, chocolate, and cocoa, and all the etceteras, like raisins, pepper, curry powder. Further, we decided to make our rations experimental, so each of us had the 3 foods in varying proportion. I was all for increasing the fats because I am certain that every ounce of fat one can digest is worth more to one in heat and energy than an ounce of any other food. We therefore started with the following daily ration each. Birdie had 12 ozs of pemmican, 20 ozs of biscuit and no butter = 32 ozs. Cherry had 12 ozs pemmican, 20 ozs of biscuit and no butter. I had 12 ozs of pemmican, 12 ozs of biscuit and 8 ozs of butter. At the end of a week Cherry felt the need of more fat, but he increased his biscuit and for a time found that it did something towards satisfying him. We all kept to our 32 ozs of food per day, and when we settled down to a proper proportion we found we were eating all that we wanted and literally craved for nothing. We not one of us missed sugar, but I found I couldn't eat my extra half pound of butter a day — the most I could eat was 4 ozs. So Cherry and I made a bargain and I took 4 ozs of his superfluous biscuit and gave him 4 ozs of my butter. This made for each of us a perfect diet — 4 ozs of butter, 12 ozs pemmican, and 16 ozs of biscuit = 32 ozs per day each. Birdie continued with his 16 ozs of biscuit and 16 ozs of pemmican, but he rarely ate more than 12 or 13 ozs of pemmican. The best diet was undoubtedly the one Cherry and I hit on, for Birdie allowed afterwards that he could have eaten butter when he couldn't eat more pemmican — but he wouldn't take it when we offered it to him. We rarely eat more than 2 or 3 ozs of our 4 ozs allowance of butter per day. Our ration, Cherry's and mine, compared with Birdie's gave 7345.6 foot pounds of energy instead of only [   ][3]. Towards evening the wind abated considerably but came on again during the night with snow and violent gusts, increasing at times to force 10. We couldn't march. The min. temp was −7.6°.

Wed 12 July   We had to remain in our bags all day. Wind force 10 from S.W. with much drift. Temp. up to +2.9° again in the morning. Towards night the wind began to drop. Bowers turned his bag for the first time from hair outside to hair inside.

Thurs 13 July   We dug out our sledges and tent, everything being deeply buried in snow. We got off a really good day's march on a decent surface 7½ miles in 7½ hours. During our march, in our efforts to avoid the pressure ridges on our right, we got too high up on to the land ice surfaces of Mount Terror and were held up by a very wide crevasse covered with a rotten, sunken looking lid which we caught sight of in a glint of moonlight just in time to avoid. We turned down its side and found it was one of many for we had run into a disturbed part of the névé running over a low mound of rock. We made east again and got on to the safer area below between the land ice and the pressure ridges. We camped at 8 p.m. There was a striking rosy glow of daylight in the north today behind the slopes of Terror. Min. temp. for last night −22.2° and by the evening it was −28.6°.

Fri 14 July   We made 5⅓ miles today, a good morning march, but the afternoon march was cut short by a complete failure of light. After lunch we found we had again gone too far east and had run right into one of the big pressure ridges again. We turned north and soon encountered more crevasses, but by zigzagging and sounding with a pole we got away from them. In the bad and doubtful and absent light in which we were travelling continually it was most difficult to keep the middle way between the traps which lay on either side of us, or when we found ourselves caught in one of them, to know which it actually was. In daylight, in the *Discovery* days, we used to have but little difficulty here, but we were never here in the dark. It seemed in the fogs and darkness which lit our way now as though the pressure ridges had encroached on the land ice everywhere. The min. temp. for the night −35°. At 8.30 a.m. it was −17.4° and in the evening −24.6°.

Sat 15 July   The min. temp. for the night −34.5°, but at 10.30 a.m. it was −19.2° with a breeze from S.S.W. We got a clear view this morning after a short uphill 3 miles pull over very deep cut hard sastrugi, we reached the shelf of moraine where we had decided to build our stone hut. Rounding the lower end of this moraine we found ourselves in the Knoll Gap and pitched our tent in a large open smooth snow hollow, hard and wind swept on the surface, but in places not deeply cut. This camp was the last one on our outward journey for we had at last reached the place where our work lay. The camp lay about 150 yards below the ridge where we proposed to build our stone hut. We had originally intended building on the Adélie Penguin rookery, but our time had been so taken up in getting here and our supply of paraffin so short, that we decided it was better to build at once and as near to the Emperor Penguin rookery as possible. In the Adélie Penguin rookery we should have been sheltered from blizzards, but here we should be right in their track — only 5 miles nearer our work. On the ridge top above this snow hollow where we camped was a low rough mass of rock in situ with a quantity of loose rock masses lying around and erratics of various kinds, some granite, some hard basalt, and some crumbly lava. There was also a lot of coarse gravel and plenty of hard snow which could be cut into slabs like paving stones.

So here we had all the material we wanted for building, and we chose a spot 6 or 8 yards on the lee side of the edge of the ridge, a position which we thought would be out of the actual wind force, but which we eventually found was all the more dangerous for that reason, as it was right in the spot where the upward suction was to be at its greatest.

At lunch time 4.15 p.m. we still had a wind of force 4 at −13° which we believed to be due to a more or less constant flow of air down the slopes of Terror. From this spot where we built our stone hut we had a magnificent outlook. To the E. and N.E. we looked over the whole of the Great Ice Barrier and Ross Sea, below us about 800 ft below was the whole range of great pressure ridges where the Barrier is moving along the shores of Ross Island. To the south we looked over the Barrier to White Island and the Bluff, and to the west we had the summit crater and all the slopes of Mount Terror, pile upon pile of snow and rock and glacier. Facing the door was the Knoll, a small extinct crater with the daylight in the north at noon behind it. Ross Sea was completely frozen over as far as we could see in the moonlight.

Sun 16 July   Fifth Sunday after Trinity and an anniversary not easily forgotten[4]. We began the building of our stone hut, and I called it Oriana Hut, and the ridge on which it is built Oriana Ridge. We worked as long as there was light enough at midday and by the waning moon. Bowers and I collected the rocks and Cherry did the building. Then we piled up the out sides with snow slabs and gravel, two or three layers thick. We had a pick and a shovel with us. The hut was 8 feet across and about 12 feet long. The weather wall was a foot higher than the lee, and the entrance, in the lee wall, just allowed us to crawl in on all fours. We laid our 9 foot sledge across the walls as a ridge beam to support a strong canvas roof which had been prepared beforehand with long valance to build in all round.

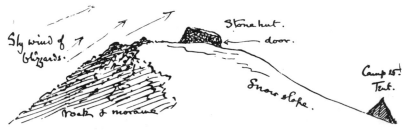

Sketch of the location of the stone hut at Cape Crozier

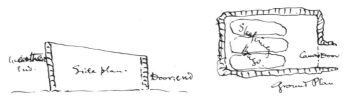

Profile and plan of the hut

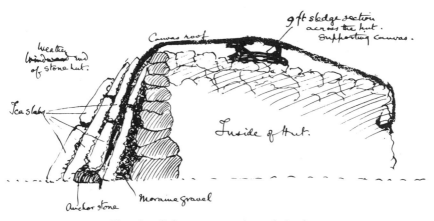

Sketch of the construction of the hut

Mon 17 July We continued building the hut and finished all but putting on the roof and door for which it was blowing too hard. Temp. −23.3° only.

Tues 18 July Temp. −27.3° and blowing too hard to finish the hut or do anything else. We therefore lay in our bags and were very cold for want of exercise.

Wed 19 July Turned out to be a calm day so we decided to make an effort to reach the Emperor Penguins' rookery and get some blubber for our stove, as our last can of oil but one was running very low and we couldn't afford to touch the last can, which we should want for the journey home. We made an early start and were away by 9.30 a.m. before any appearance of dawn. We took an alpine rope, ice axes, harnesses and skinning tools and a sledge. We had about a mile down snow slopes to the first pressure ridge, and our intention

93

was to keep right in under the land ice cliffs, which are now very much more extensive than they were 10 years ago, and then on under the actual rock cliffs which had always been the best way before in the *Discovery* days. Somehow, in the bad light, we got down by a slope which led us into a valley between the first two pressure ridges and we afterwards found it impossible to get back in under the land ice cliffs. We tried again and again to work our way in to the left towards the spot where the land ice cliffs joined the rock cliffs, but though we made considerable headway between whiles along snow slopes and ridges by crossing the least tumbled parts of the pressure, yet we could not get a path through and the short midday light, which was absolutely essential here, was fast running out. We were working in a maze of crevasses all the while, roped together of course, and bridging the crevasses we couldn't cross with our sledge. It was quite exciting work but it grew very much more exciting as the light got worse and worse. All round us was a chaotic mass of tumbled blocks of ice with crevasses everywhere either gaping wide or else bridged with snow drift. We tried one opening after another and found them all impossible until at last we were faced by a small mountain of ice blocks shutting off any further progress and about 60ft high all round us, we being in a drift-filled valley at the bottom. Here we had the mortification of hearing the distant cries of some of the Emperor Penguins echoed to us from the rock cliffs on our left. We were still, however, as far from the sea ice and the rookery as ever and more than two thirds of our daylight had gone. So we were forced to give up any further attempt that day and with great caution and much difficulty, owing to the failing light, we retraced the steps it had taken us three hours to make, and we were all but benighted before we reached safe ground after 5 hours clambering to no purpose.

During the day a light southerly breeze had been blowing with a clear sky. The temp. had varied from $-30°$ with southwesterly wind to $-37°$, which had been the minimum in the early morning between 3 a.m. and 9.30 a.m.

Thurs 20 July We turned out at 3 a.m. and got the canvas roof of our hut, with the door piece, fixed on and built in and made safe with blocks of snow and gravel and rocks anchoring it all round. It was a fine hut and would have made a top-hole living place in any other land. All that we wanted now was blubber for our stove, and as the weather was propitious we decided on another attempt to reach the rookery. We again made an early start before daylight broke. We took the same equipment and as before tied ourselves together and

94

took crampons on our feet and ice axes. We had a tremendous clamber and reached the sea ice and the rookery. This time we found quite a different narrow snow slope off the land ice cliffs down into the first hollow. We had missed it yesterday when walking along the cliffs looking for a way down. Today we got it and it took us where we wanted to get, right in under the land ice cliffs, which show what wind can do. They are a hundred or more feet high and the whole face overhangs from having been scooped out into vast grooved hollows as though by a gigantic gouge. By following the foot of these cliffs one comes to places where the black basalt cliffs stand out of the ice at the foot of the Knoll on its eastern side and by a series of slides and climbs and scrambles between ice and rock and snow drifts one comes to where far loftier ice cliffs and rock cliffs take the pressure of the Barrier ice. It is at the foot of these cliffs, amongst rock debris and snow drifts and boulders which have fallen into the trough, that one comes to a possible path down to the sea ice. It is an extraordinary climb all the way, up and down in one cul de sac and over a wall into another. At one place we appeared again to be held up by an impasse, for we came to a spot where one of the biggest broken pressure ridges butted right into the rock cliff and the only way through was a man hole in the ice just big enough for each of us to squeeze through in turn. Here, therefore, we had to leave our sledge as it wouldn't come through. Once past this we were in a snow pit with almost vertical walls which required 15 steps to be cut before we could climb out. From this we again got into a series of drift troughs between the rock cliffs and the ice of the pressure ridges, and at last we got out on to the ice foot overlooking the sea ice, and there were the Emperor Penguins. There was a small but troublesome overhanging ice cliff of 10 or 12 ft which we could easily get down but couldn't so easily get up again. Everywhere else it rose to 20 or 30 ft and the edges corniced and overhanging. So we left Cherry on the top to help us up with the alpine rope and Birdie and I jumped down. The light was rapidly going and we had some very bad crevassed pieces of pressure ridge, as well as a lot of snow bridges and a razorback ridge with 30 or 40 ft to drop each side, to go back over on our way home. If we got benighted here we should have a nice long time to put in in the dark without food or shelter, something like 18 hours; that was why we were so careful *not* to get caught by the darkness. We rushed over the sea ice to where the Emperors were huddled together in a heap under the Barrier cliffs. There were, if anything, less than 100 instead of something like the 2,000 we found here in 1902. We saw at once that many of them had eggs on their feet because they moved along so very slowly and carefully while the others walked easily as usual. We collected 5 eggs, of which only 3 unfortunately reached camp intact, and these split when they froze on the way home. We also, seeing that there were no seals on the ice here, quickly killed and skinned 3 Emperors and carried the skins home with the blubber on for our lamp. They were exceedingly fat, and the oil burned splendidly, even better than seal blubber oil. As we hustled the penguins to take some of their eggs we noticed an extraordinary thing. There were eggs dropped here and there on the ice, and as Birdie and I were collecting those we took, we both picked up lumps of rounded dirty ice the size and shape of eggs. We picked them up first thinking they *were* eggs and put them down when we found we were mistaken. While I was skinning one of the birds I saw an Emperor walk back close to me. It came across this lump of dirty ice and immediately proceeded to tuck it in under its flap of feathers on its feet and incubate it! Seeing that there was a real egg getting cold close by I fetched it and put it down close by this bird, when it at once dropped the cold nest egg and came and incubated the real egg.

We had to hurry away after this as it was all the while getting darker and darker. Then came a difficulty in getting up the little ice cliff. I got on Birdie's back and was soon up, but Cherry and I, hauling both as hard as we could, were unable to haul Birdie up, as the rope cut into the cliff the more we pulled and did nothing else. We got him an ice axe and then by hauling on him as he cut himself steps we at last got him up. On the sea ice Bowers unfortunately stepped through a crack and got one of his feet and legs wet through. They of course froze into a solid block, but he didn't get very cold. We had a difficult clamber back in the failing light and got in when it was quite dark — back to camp. On returning we at once flensed one of the Emperor skins, and cooked our supper on the blubber lamp, which gave out a furious heat and nearly choked us altogether. We soon got black and greasy. We slept that night for the first time in our stone hut, and it was a bad night for me for I got a spirt of burning oil in my eye and it gave me great pain for a good many hours[5]. The temp. had not been below −28.3°. In the evening the wind freshened to force 6 from S.S.W.

Fri 21 July Our first night in the hut was good save for my eye. The wind rose to force 8 and had an effect on the canvas roof which we didn't quite like the look of. It made it quite taut at the same time just raising it off the supporting sledge rafter. We couldn't quite understand it and certainly didn't quite grasp the significance as we should have done. The wind dropped next day but looked unsettled. We spent the whole day in further packing every crack and crevice of our hut with snow, and we strengthened it in every way we could think of. Also put large slabs of icy snow on the canvas roof to counteract its tendency to 'lift' in the wind. We thought it was perfect when we had finished. Then we brought our tent up and pitched it under the lee of the hut quite close to the door for convenience, as we found it easier to warm the tent with our blubber lamp, concentrating the heat on wet socks and mits more than in the hut. We therefore decided to have our meals in the tent in future and to sleep in the hut until we could get seal skins to put over the canvas roof and keep the heat in. When we turned in the sky was overcast but there was no wind. We slept in the hut and my eye was very much better. The wind came on suddenly at 3 a.m. from the south and blew hard with little drift.

Sat 22 July By 6.30 a.m. it was blowing 9 to 10 with heavy drift and very violent gusts, and soon after this when Bowers looked out of the hut door he saw that the tent had been bodily blown away, legs, lining cover and all, leaving most of the gear we had left in it over night on the ground. The drift was very thick and the only thing to be done was to collect the gear and pass it all into the hut — which we quickly did. We found that two pieces of the cooker had been blown away — happily not the most essential parts, either of them. None of our finnesko had gone, happily, and a lot of mits and socks and such like small things, which might easily have followed the tent, were still there on the ground. My fur mits, however, were never seen again. Meanwhile in some inexplicable way the snowdrift was finding its way through every crack and crevice of our stone hut notwith-standing the canvas cover and sides and snow and gravel packing. It came in in such quantities that we and all our gear were soon inches deep in it, and later when the drift eased up we were layered down in fine black gravel dust which came in in the same way and smothered everything. We tried plugging the cracks with socks and mits but it wasn't any use. The wind was so strong and blew so hard over the roof of our hut that it sucked the canvas up and tried to lift it right off. It was sucked up as tight as a drum about 6 inches off the sledge which was supposed to be supporting it. This suction produced a sort of vacuum inside the hut and thus the snow dust and gravel dust was sucked in from outside, actually the stuff we had used for packing. So long as the large ice slabs, which we had put on the roof, remained there, there was no flapping or friction of the canvas on the stones, but before the next night was over they were all blown off. However, as the storm continued all day, we decided to cook a meal on the blubber stove. One comfort we felt was that we had 3 penguin skins for this purpose which would last us some days. Alas — we got the stove going, and after a few efforts it had nearly boiled the water for our meal when suddenly the heat unsoldered a feed pipe which ought to have been brazed, and the stove was at once useless. We poured the oil off into this for use on the way home in case of necessity, and then began to consider matters in the light of having only one remaining can of oil for the journey home and no tent. We still had a little oil in our last can but one — but it was obviously necessary for us to avoid touching our last can — and as soon as we are forced to we must go. It looked like coming to this very soon, for it would be very difficult now to improvise a stove without materials. As for a tent we believed we could improvise a shelter good enough to get home under with the canvas roof of our hut. Lying in our bags ruminating over these things in the hut, we were sopping wet and getting wetter every time we had to get in and out of our bags for anything, there was so much snow drift in the hut, but we were not cold, thanks to the rise of temperature with this blizzard. We finished our meal on the primus lamp when the blubber lamp gave out — and this was our last meal for a good many hours as it happened. The disappearance of the tent was a great surprise to us, for we had got a perfect spread and we had taken every possible precaution against wind. The valance was first buried in snow and then we added 4 or 5 rocks to the snow in each bay — and as a final precaution we slung a very heavy canvas tank of provisions and extra gear, so heavy that Birdie and I had to lift it together, on to the skirt. We can only think that the same lift acted on it and carried it away. Anyhow it was gone and out of sight, and we couldn't look for it till the blizzard stopped and a little daylight gave us a chance. Worse was to happen before this happened however.

Sun 23 July   Sixth Sunday after Trinity and quite the funniest birthday I have ever spent. The wind was terrific. It blew almost continuously with storm force — there were slight lulls occasionally followed by squalls of very great violence, and at about' noon the canvas roof of the hut carried away and we were left lying exposed in our sleeping bags without a tent or a roof. The storm continued all day with unabated vigour. There was no choice for us now — we had decided this before in the event of our roof being blown off — we had to remain lying there in our bags till the blizzard stopped. We had had two days of it, but here at Cape Crozier Royds in 1902 was laid up with his party for 5 days — and I with mine for 8 days out of 11. So we could only hope that this was not going to continue quite so long. If it did our best chance was to allow the snow drift to cover us up, which it was doing already, in order that we might at any rate keep warm. We could always eat biscuit and cold pemmican in our bags and we all had biscuit in our pockets.

When the roof went Birdie and I were both out of our bags, for we were trying to stop the flap and the chafe of the canvas which began when the snow blocks were blown off the roof. The weakest spot was where the door came, but we had anchored it with very large stones. These stones the wind acting on the canvas joggled about like so much gravel and they gradually shifted out of place. We did all we could to jam them tight, but to no purpose, for while we were still at it the canvas ripped out all along the lee end of the hut with a noise like a battery of guns going off. In a second the canvas was ripped in about 10 places and it flapped to bits from end to end in a few minutes — leaving a ragged, flapping end attached to the weather wall which then went on bang, bang, banging for hours till the wind eventually dropped. The noise was most distressing, and we hardly noticed the rocks that fell in, or that the sledge was at once flapped off and fell in also across our three bags. We were at once in a perfect smother of drift when the canvas carried away, and Birdie and I bolted into our bags taking an enormous amount of snow in with our clothes to thaw out at leisure. We were not really so much disturbed as we might have thought, and we had time to think out a plan for getting home again now without our tent — in case we couldn't find it — and without the canvas roof of the hut which had gone down wind in shreds the size of a pocket handkerchief. We still had the floor cloth of the tent, and this we were lying on so it couldn't blow away. We could build a snow hut each night on the way home and put this over the top; or we could always dig a burrow in the Barrier big enough for the 3 of us, and make a very good roof with canvas flush with the surface — if there was wind it couldn't then be blown away. We had no doubts about getting back so long as this blizzard didn't last till we were all stiffened with the cold in our bags. The storm continued all day and on until midnight unabated.

Mon 24 July   At midnight the wind began to get more squally and dropped in force from 11 to 9 with short lulls. At 6.30 a.m. it had dropped to force 2. At 10 a.m. it was about force 3 and we waited for the moment when there would be light enough to go and look for the tent. While it was still dark we managed to cook up some pemmican, our first warm food for 48 hours. We did it by sitting up in our bags and drawing the floor cloth over our heads to protect the primus. When at last dawn came along it was by no means reassuring to see that in the south there was evidently a lot more blizzard still to come, so we lost no time in getting away down the slopes to leeward where our tent must have gone. Everywhere we found shreds of the roof canvas, and about a quarter of a mile round a corner of the hill Birdie found the tent — poles, cover and all hardly damaged. One of the poles was torn out of the cap, but that was nothing and soon remedied. We quickly brought it back, pitched it lower down the slope where it would be more sheltered, and then carried down our bags and the cooker and primus, and all the essentials and once more felt we were prepared for any amount of blizzard. It looked as though we shouldn't have long to wait. We then discussed our position and came to the conclusion that with one can of oil only we couldn't remain here and improvise a blubber stove for tent work, seeing, too, how very difficult the way down to the sea ice had become.
It was disappointing to have come all this way and to have done nothing worth mentioning of our work with the Emperor Penguins, but to remain now had become a practical impossibility. We had to own ourselves defeated by the Cape Crozier weather and by the darkness, which was really our greatest difficulty all through. We decided to put everything in order here today, to depot all that we could possibly leave that would be useful for [our] next visit here in the hut, and tomorrow to start back. We left the box of pickling solutions and all the apparatus I had brought for the penguin work. We also left a sledge, a pick, some bamboos, a variety of odd clothes and a down sleeping bag which was so solid with ice that nothing could be done with it. This was Cherry's. I also wrote and tied a note to the handle of the pick where it couldn't be missed, in a match box.

Tues 25 July   Stiff breeze from the south. Temperature −15.3° with thick weather coming up, so we quickly made our final arrangements, and started off down towards the pressure ridges

100

again with one sledge and our camp kit. We found that our sleeping bags were now so stiff with ice that rolling them up besides being difficult actually was a danger for the leather broke instead of bending. We therefore from now onwards laid them one on the other full length flat on the sledge. They were really about as bad as they could be and we were getting very little sleep in them at night. Cherry felt this most. Birdie slept most — I slept very little but didn't feel the loss of it much. I still kept my down lining in the bag though it was flat as sheet tin and about as warm and soft, but it held my reindeer skin bag together, which otherwise would have come in two pieces across the middle. My bag was broken at both ends as well as two big rents across the middle and the head eventually shrunk so hard and so immovably that I couldn't close the flap over.

The journey home from here was by far the coldest experience I have ever had and by the time we got back to Hut Point we were all so short of sleep that not a meal passed without our having to wake each other up for fear of spilling the pemmican or falling into the cooker or the lamp. Well — we made a start home today but we hadn't gone more than a mile when it came on to blow so hard in our teeth that we had to camp again. We were then amongst very big, hard, icy sastrugi which made the tent pitching very difficult, but by anchoring the flap to a heavy case of biscuit we made it safe, and Birdie also tied himself in his sleeping bag to the tent poles, quite determined that if it blew away he would go with it and bring it back again. The gale continued and freshened to force 9, and lasted all night. Temperature during the day was from −15.3° to −17° and the whole sky was overcast.

Wed 26 July   The short midday light was all but gone when the wind dropped enough to allow us to start again. Leaving at 2 p.m. we then made 4½ miles in 3½ hours and once again got in amongst crevasses in the dark on smooth wind swept ice. We continued, however, feeling our way along by always keeping off hard ice slopes where the crevasses are bridged with rotten snow, and by keeping as far as possible to the crusty deeper snow which characterises the hollows of the pressure ridges which I believed we had again got foul of in the dark. We had no light and no landmarks to guide us except vague silhouetted slopes ahead which might be yards away from us or might be miles, but whose character it was impossible to judge. We travelled as much by ear and by the feel of the snow and ice under foot as by sight. We got to know very fairly well whether we could go ahead confidently or whether we were likely to fall through a lid. Probably my own feelings were quickened more than the others as I was on a long trace leading by about 10 or 15 ft ahead. The sky cleared when the wind fell and we had the temperature dropping from −21.5° at 11 a.m. to −45° at 9 p.m. We made our night camp among some pressure ridges on soft snow in one of the hollows.

Thurs 27 July We got away with the approach of daylight and found that we had camped, as we thought, right in amongst the larger pressure ridges and somehow, without crossing any large ridge yesterday, we had crossed several smaller ones safely in the dark and had then come on for a considerable distance between two very high ridges. Ahead of us was a safe and clear road along this hollow to the Barrier in the south, but our direction was S.W. and as the pressure ridge crevasses were so thinly bridged everywhere we hoped that by continuing along this hollow for a bit we might find some spot on the right where it would be low enough to cross more easily and so get on to the land ice again. We found no such dip, however, and after going some distance out of our course decided we must cross the ridge where we were. The whole thing was crevassed and one was so broad that we were all on the bridge with our sledge at once in crossing it. The next one I discovered by getting a leg down, but as I shouted a warning to the others I saw Birdie disappear down it, happily hanging by his sledge harness. He was quite helpless but with the alpine rope we gave him a loop for his foot and soon hauled him up again. After this we got on to better ground and soon reached safe land ice. We next got on to a very long upward slope and made good going till we had to camp, having covered 7½ miles in the day. The temp. varied from −45° to −47° during the day, but the weather was calm and what was more to the point, was clear enough latterly for us to see something of where we were going to in the dark.

Fri 28 July We got away before daybreak and found ourselves still on the upward slope of a very long gradient with a gentle breeze of cold air flowing down it. The Bastion Crater was on our right with the conical hill surmounting it, a landmark visible from Observation Hill. We went on and on up this slope until at last we found ourselves in a calm on the divide with a magnificent and encouraging view of the Western Range of mountains, Mount Discovery and Hut Point Peninsula, all showing like ghosts in the dim light. We then knew we were well over Terror Point and getting out of the blizzard area into the colder one. The surface all up this slope was good going. Over the divide we went down hill with the air stream on our backs. These local streams of cold air are typical of calm weather in this country; they occur on slopes. We soon got down on to the smooth soft Barrier surface again, our feet and the sledge runners both sinking in some inches. Subsidences also became frequent again.

Bright, fine weather — Erebus and Terror top showing all day. Temp. ranged from −47.2° in the morning to −38° in the evening. We made 6¾ miles in the day. Our sleeping bags are so wanting in any sort of attraction that we now cut our stay in them as short as possible, turning out at 5.30 a.m. and turning in about 10 p.m. We have dim daylight from about 10.30 a.m. when dawn begins, to about 2.30 or 3 p.m., but the sun is still so far from getting above the horizon at noon that it is only a sort of dim twilight all those hours at the best. The sun will not show himself at all above the horizon until August 23rd. However, even this twilight is a great blessing now as we have no moon, and we arrange our marches to make use of it all before camping for lunch. After lunch we march in the dark. Our hands give us more pain with the cold than any other part — our feet are generally warm, but our hands are often dreadfully cold all night, soaking wet in wet mits the skin is sodden like a washerwoman's hands, and in this condition they get very easily frozen on turning out. Cherry's finger tips are all pretty badly blistered. I have only one bad thumb which blistered early and is now broken and very sore.

Sat 29 July   Made 6½ miles. Subsidences very frequent and at lunch the whole tent and the cook[er] and contents dropped suddenly with a bump and with so long and so loud a reverberation all round that we all stood and listened for some minutes. No wind today. Temp. from −42° in the morning to −45.3° in the evening.

Sun 30 July   Seventh Sunday after Trinity. We had fine weather again and did 7½ miles in the day. Dawn on the horizon in the east is deep carmine running in a broad band along the Barrier horizon under a green sky. The temp. was low, −55.3° in the morning; −63.2° to −61.8° in the evening — pretty severe in our solid frozen wet sleeping bags. We have very poor nights and very little sleep. Fog banks were forming all along the Castle Rock ridge where the cold barrier air met the warmer air on the sea ice side of it.

Mon 31 July   Turned out soon after 5 a.m. Calm weather again and we made 5½ miles in 5½ hours before stopping to lunch at the edge of the Barrier, 1½ miles to the S. of Pram Point ridges. Here we ran down a drift slope on to the sea ice about 12 ft below. Temp. −57° at this point where we lunched. After lunch −43°, and when we reached the hut at Hut Point it was up to −27°. We took very little time covering the 3½ miles between the Barrier edge and Hut Point. Here we remained the night and decided to be up at 3 a.m. and get back to Cape Evans by dinner time.

Tues 1 Aug   We slept where we sat in our tent which we pitched inside the hut, which was very cold indeed, but the tent we warmed up with a primus lamp as there was lots of oil here. We were in our bags for 3 hours, but at 3 a.m. it was blowing too hard to start and we remained dozing where we were until 10 o'clock. Then we had a meal and got away at 11 a.m. when the wind dropped. We marched on as long as there was any light and then camped for lunch off the Glacier Tongue. Then the new moon rose and we got away and were in at Cape Evans on the 36th day of our absence at about 10 o'clock in the evening. It was a great comfort to get off one's sopping and frozen garments and to turn into bed after a good supper of cocoa and bread and jam. We were pretty tired for want of sleep.

Wed 2 Aug   Warm bath. Hair cut, shave — some extra sleep — and wandered round enjoying warmth and dry clothes. Feet, hands, nose and mouth all tingling and rather swollen and sore now that the reaction has set in. We have none of us lost much weight. Cherry only 1 lb, Birdie and I 3½ lbs each.

Thurs 3 Aug   Enjoying comfort — with an enormous appetite. During our absence they have had two excitements. One was that Atkinson was very nearly lost in a blizzard. He was out 5 hours completely lost, and the whole party out in various directions looking for him. He came in eventually with his hands very badly frost bitten. The other excitement was that one of the ponies very nearly died of colic.

EPILOGUE

It is of interest to compare Wilson's low-key account of this most hair-raising of all Antarctic field trips with Cherry-Garrard's dramatic narrative of events published as a chapter in his classic account of the expedition *The worst journey in the world*. 'Antarctic exploration,' he concluded, 'is seldom as bad as you imagine, seldom as bad as it sounds. But this journey beggared our language; no words could express its horror.' Despite their profound exhaustion, Wilson, Bowers and Cherry-Garrard rapidly recovered their spirits and soon became involved in the detailed work of preparing for the

Pole journey. As for the three Emperor Penguin eggs, for which 'three human lives had been risked three hundred times a day', they were to have a chequered history. On the return of the *Terra Nova* to England the eggs were handed over to the British Museum (Natural History) in London. The embryos were then entrusted to a Cambridge don who died in 1915 before he had completed his research. They were then passed to a professor in Edinburgh who kept them until his death in 1933 without having reached any final conclusions. By 1934, when a third expert had concluded that the development of the penguin embryo could not shed any new light on the problem of the ancestry of birds, several other similar collections had been described.

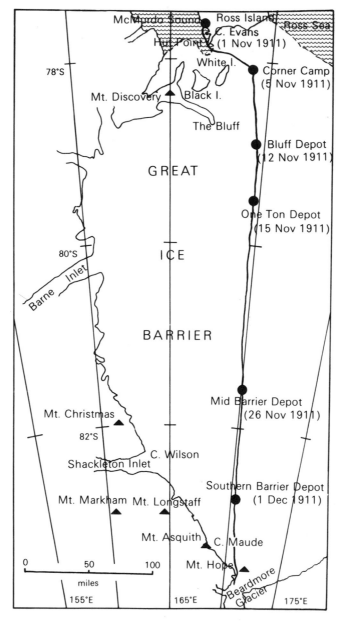

Map of the Pole Party's route from Ross Island

4

SLEDGE JOURNEY
TO THE SOUTH POLE

PROLOGUE

AFTER the rigours and excitements of the winter journey to Cape
Crozier the weeks remaining at Cape Evans before the
departure for the Pole were spent by Wilson in necessary, if
uneventful, pursuits. There were the ponies to be exercised and
curbed — 'a rotten lot' as he described them, selected somewhat
haphazardly from the Vladivostok region of Russia by Cecil Meares
and trained in the Antarctic by Captain Oates, who tended to get the
blame for their shortcomings and eccentricities. Scott's anxieties for
their behaviour are understandable since he was relying on them to
haul the bulk of his provisions to the foot of the Beardmore Glacier,
here to be shot, butchered and the meat cached for the returning
parties. As chief of the scientific staff, Wilson was kept busy
supervising an ambitious scientific programme and helping the
youthful scientists working under him, discussing with them their
specialities and generally acting as father-confessor. Writing to Mrs
Wilson in praise of her husband, Scott wrote: 'Even to you I have no
words to express all that he has been to one, and to the expedition,
the wisest of counsellors, the pleasantest of companions, the loyalest
of friends.'

As the time for departure grew nearer, Wilson became much
concerned with the likelihood that if he were to be chosen for the
Pole he would not be back at the hut in time to catch *Terra Nova*
before she sailed for New Zealand sometime in March 1912. But
writing up his diary he reflected philosophically: 'I don't see any
other course open to me than to carry through the job I came here
for, which was in the main this sledge journey for the Pole.
"L'homme propose, mais le Bon Dieu dispose" is an honest creed,
and in this case l'homme hasn't decided to do anything from first to
last that he wasn't convinced would be approved by his infinitely
better half, and le Bon Dieu will do the rest.'

Despite all these responsibilities Wilson still managed to find time
for his sketches. By October he had completed a total of 118 of them

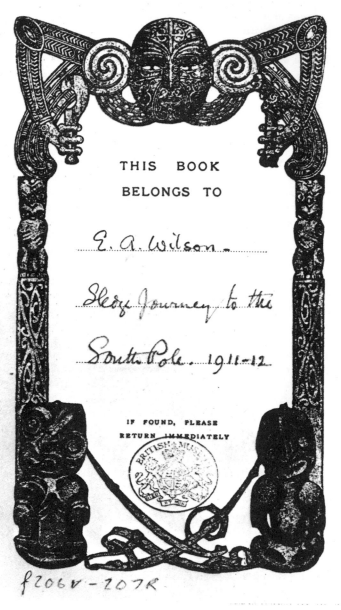

THIS BOOK
BELONGS TO

E. A. Wilson —

Sledge Journey to the

South Pole. 1911-12

IF FOUND, PLEASE
RETURN IMMEDIATELY

f 206v - 207R.

Title page of Wilson's sledging diary

and was planning arrangements for their exhibition in London the following year alongside the photographs of the expedition's camera artist, Herbert Ponting, who was planning to return to England immediately.

By the last week of October all was ready for departure. On the 24th the two motor-sledges, driven respectively by Bernard Day, the mechanic, and Stoker Bill Lashly assisted by his fellow seaman Frederick Hooper, and under the leadership of Lieutenant Teddy Evans, set off for Hut Point and the Barrier. Pulling a ton-and-a-half of stores apiece the job of the Wolseley motor-sledges was to move the bulk stores as close as possible to the Beardmore Glacier, the route to the polar plateau discovered by Ernest Shackleton in 1908. On previous trial runs the motors had revealed numerous mechanical weaknesses and were constantly breaking down. Eager as he was to see them succeed, Scott privately felt despondent. On this occasion, and after much tinkering, they did achieve the Barrier, where overheating of big-ends led to their eventual breakdown.

On 31 October, Wilson wrote final letters to catch the relief ship and finished off his journal to accompany them. Based on his rough field notes the journal had been intended to serve in part as a newsletter for the family. The sledging diary, which we reproduce here in its entirety, is very different. Cherry-Garrard, who drew on it extensively in his own narrative, describes it as 'that of an artist watching the clouds and mountains, of a scientist observing ice and rock and snow, of a doctor, and above all of a man with good judgement. You will understand that the thing which really interested him in this journey was the acquisition of knowledge. It is a restrained, and for the most part a simple, record of facts. There is seldom any comment, and when there is you feel that, for this very reason, it carries more weight.'

The sledge diary opens on 1 November 1911 with the departure from Cape Evans of the main party en route for Hut Point from where, after a brief rest, they were to cross over the sea ice and on to the Barrier.

DIARY

Wed 1 Nov   Left Cape Evans. 10 horses. 10 men and loaded with some Hut Point provisions — about 11 a.m. Arrived about 3.30 p.m.

Thurs 2-Fri 3 Nov   Remained at Hut Point all day till 9 p.m.

when Owner, Cherry and I with Snippetts, Michael and Nobby followed after an advanced unit composed of Atkinson, Wright and Keohane with Jehu, Chinaman and James Pigg — the three weakest ponies. They started 2 hours in advance of us. We caught them up at Safety Camp 6 miles — where we lunched. The other 4 ponies Christopher, Bones, Victor and Snatcher came right through without stopping for lunch about an hour after us. With them were Oates, Crean, Bowers and Evans. After lunch we and the advanced unit went on again 6 miles where we all camped for the day. We were turned in about 6 a.m. Wind which has been cold from N. all night with a good deal of drift at Hut Pt. and C. Armitage now dropped. At 1 p.m. the horses were fed. Turned out to a glorious hot day for half an hour. So hot left the tent door open. Slept all night in which was really day.

Fri 3-Sat 4 Nov   We had breakfast at 8 p.m. Did 5½ miles by lunch and 4½ afterwards. All the horses went well. Surface pretty soft well up to

at every step the long hair spreading out round the fetlock on the snow. Here and there a crust hard enough to hold them up. Gradually overcast towards morning and soon after turning in some wind and a little drift. I am cook this week.

Sat 4-Sun 5 Nov   Overcast day but warm and calm — turned in all day except at horse feed time 1 p.m. All day yesternight we were coming across derelict cans of petrol, lubricating oil and eventually Day's derelict car itself, with the 'big end' broken — part of the piston of one of the 4 cylinders. Also, a huge pit at another camp where the party had been blizzarded and the other car dug out. £1,000 by the wayside! We had 9 miles to go before reaching Corner Camp. Did 5 before lunch, 4 after and about a mile on the home side of Corner Camp 6 horses got their feet into crevasse cracks at one place. It must be a regular honeycomb on a line rather E. of between C. Crozier and the Bluff. At Corner Camp we spent another fine warm sunny day. Can see what looks very like the other motor about 4 miles S. From here we go due south. We load up a lot of stuff here from the depot. Nobby's load amounts to about 650 lbs.

Sun 5-Tues 7 Nov   We had a good march the night thro' in 2 pieces, but shortly before camping for the following day it began to

110

blow and from being overcast to the E. and W. only, became overcast everywhere, so we prepared for a blizz and it came.

We therefore lay in all Monday — and Monday night and all Tuesday, turning out at intervals to see to the horses which includes feeding, rebuilding parts of their walls which they knock down, digging out snow drift accumulations, and knocking the lumps off their hoofs. My beast is comfy enough, feeds well and eats plenty snow and takes interest all round. Some of the others look rather wretched. Meares and Demitri with the dog team came up with us early this morning and camped a quarter mile off. I have spent most of my two days and a night asleep. My capacity that way is enormous just now. On Sunday read H.C.[1] and a chapter. On our Sunday night's march we passed the second car also derelict, and 2 sledge loads of gear, petrol, forage and 1 complete unit of man food for a week. This is buried in snow under after-end of car. Also one bag seal pemmican and 1 can of oil on the car. This car was also a big end break. And it was left about 1¼ miles S. of Corner Camp.

Tues 7-Wed 8 Nov   On Tuesday night the wind dropped a bit though it remained overcast — but about midnight we got under way and went along in two pieces for 10 miles as far as two cairns, one of which is Blossom's[2] grave — the northernmost — and has a wire and a bit of bunting on it, black. Meares and Demitri and the dogs came up with us here again and I went over and had a talk with him. All's well. The weather gradually cleared all night and wonderful wreaths of cir. str.[3] rose vertically under the sun east to flow S.W. in long swathes. We had a glorious sun to camp in and a glorious day for our night's rest all Wednesday.

Wed 8-Fri 10 Nov   At 10 p.m. we again got away — still in 3 detachments, all horses going strong and well, and by 8 a.m. on Thursday we were camping again in sunshine — light breeze. Wind came up and we made two marches in blind white snowy weather against mod. breeze. We passed a last year's cairn about a mile on our left. We following Teddie Evans' tracks are too much to the west. We camped in the same weather and spent the day in sleep.

Fri 10-Sat 11 Nov   Turned out at 9 p.m. to the same weather again on Friday evening and made 9½ miles in 2 pieces over an abominable surface. In the first 5 miles we passed a cairn made recently by Evans and an old cairn of last year's about a quarter mile N. of our lunch camp where we again built a cairn. There were no cairns between lunch and camping for the day's sleep, but tomorrow's

Some of his companions sketched by Wilson on his way to the Pole

march ought to bring us to Blucher's[4] cairn in about 2 miles and to the Bluff depot in 5 to 6 miles from where we are today Sat. Nov. 11.

Sat 11-Sun 12 Nov   We could see the mainland to the W. today beyond the Bluff. Much deep snow has fallen everywhere. We came on 10 miles in 2 stretches again during the night of Saturday to Sunday. Chinaman going very groggy indeed — Jehu rather surprisingly well. We passed Blucher's cairn with some wires and a bit of black bunting and then the Bluff depot cairn — no depot but a flag staff with a tin biscuit box and cairn.

Here was one of Teddie Evans' camps. The surface was of very deep soft snow and snow fell in light airs at various times of the day — much coming from the N.E.

Sun 12-Mon 13 Nov   Camped Sunday morning thus

Cherry took over the cooking at the supper meal. Snowed most of the day. Read H.C. and other things. Still snowing all next night when we did 6 miles in the first piece passing a night camp cairn of Evans' at about 2 miles. Surface very deep soft snow. Parts of fine parhelion occasionally. Also passed some of last year's horse walls at about 5 miles. After lunch we did 4 miles passing a cairn on the way and camped for Monday in hot sunshine which blessed us for an hour or two. Temp. in tent +51° quite warm and comfy. Surface 6 to 8 inches deep in soft flocculent snow like swan's down — large plate crystals. The clouds and light and shade effects have been very beautiful.

Mon 13-Tues 14 Nov   Snowing off and on all day and still snowing these crystal plates all the night. We got in 2 marches, 6 miles and 4 miles, through very heavy surface passing a good cairn at about 5 miles and another at 8 or 9 miles, and camped this Tuesday morning.

One of two sketches showing colours of parhelion

Tues 14-Wed 15 Nov   We have another yet about a mile or two ahead of us. Camp arrangements of horse walls on the 13th was irregular.

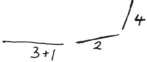

Camping arrangement for 14th was,

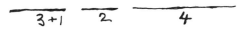

Fine hot sunshine after the snowy night. Dog teams both up with us. Surface 6 inches deep with swan's down fluff. Clouds have been radiating S.E. and N.W. all day. Got in another march in 2 pieces during the night and reached One Ton Depot on Wednesday morning. We passed a newly made 10 ft cairn of Evans' at about the 3rd mile of the first march of 6 miles and about a mile further on an old squat cairn of last year's. After lunch we had only a mile to go and we reached One Ton Depot with its main cairn and flagstaff and 2 sledges.

One Ton Depot We dug for oats but found none. There is a regular hill of drift from a W.S.W. wind.

Our horse walls here are arranged

We had a magnificent show of solar corona 3-4 and even part of a 5th ring showing. They were best seen through snow goggles as the glare was excessive — but though there were more rings at mid night, the rings were more perfect as the sun got higher. The cause of them was wreaths of fog about 20 to 30 ft over our heads moving rapidly

115

and visibly across beneath the sun. The main colour was in the orange red rings which were very distinct but in the two parts below the horizon emerald green was vivid and brilliant as well. The blue and violet was more of a blue grey than distinctive colour. We rested the ponies at One Ton Depot for a day and a night and a day. So on the night of Nov 15 we made no march. The dogs are here resting also. We find all our provisions intact and a note from Evans showing him to be gaining on us and now 5 days ahead. The minimum temp. for the winter as given by thermometer left here was −72°F. Our present temp. is −15, but was up to +7 yesterday.

Thurs 16-Fri 17 Nov   We have a cold Sly [Southerly] wind blowing all day with snow crystals driving and now and again beautiful parhelion in a blue glittering sky. We see land which must be Mt. Markham from here. Also Evans' next cairn due south. Nobby's load from here is made up to about 580 lbs, 8 biscuit boxes + 100 lbs oil + standing wts. We have depoted some seal's liver here. Slept pretty well all Thursday − temp. too low to be very pleasant outside. The 3 crocky ponies got away about 7 p.m. and we with the remainder all together at about 10.15 p.m. and marched 7½ miles geog. when we camped and had lunch. We passed on the way 2 cairns made by Teddie Evans − the first at about 2½ miles − the next at 5½. We made twin cairns at our lunch camp. There was iridescence in the flocculent looking sheet of small cir. clouds − which was very uniformly thin over a large area − thicker over another part. The iridescence showed only where the sheet was thinner − and here it was universal within the 22° circlet right up to the sun's brilliance. After lunch we did six miles more and camped with a very cold breeze − surface drift and ice crystals in the air − with temp. down to −18°. It was therefore pretty cold.

Fri 17-Sun 19 Nov   A fine parhelion − all coloured except the horizontal band and the mock sun beneath on the horizon. Surface much better than last march two nights ago. We passed a night camp cairn of Evans' at 4 miles from our lunch camp. We made horse walls thus

and the 3 + 1 wall fell down in the middle making a big breach. We had a sunny day but some Sly breeze and drifting crystals. We got away at 10.30 p.m. and lunched at 2.30 a.m. passing on the way a lunch cairn of Evans' − and a night camp cairn also of his. His tracks are covered by an inch of snow crystals but quite visible. The surface today is better to look at — more of the original to see but the runners seem heavy. The horses were rather slower today — much mirage, clearer sky, clouds only to the west but land visible. Can see Erebus and Terror now and then — looking small. Temp. at lunch camp −20°F. but no wind. After lunch we made another 6 miles with deep soft surface but the ponies went well. We passed an Evans' cairn at the end of the first mile — and then no other — but today where we camp there is another cairn close ahead of us. Our horse walls at this camp are

$$3+1 \qquad 4 \qquad 2$$

Sun 19-Mon 20 Nov   Read H.C. and other things. This is Sunday and here we are dumping a sack of oats. Jehu and Chinaman are to last only 4 more days. We had a fine day for sleep — clear, sunny and no wind — and got off about 10 p.m. The crock brigade of 3 ponies got off a couple of hours earlier. We passed a cairn of Evans' in a quarter of a mile and the next which was a lunch camp dated Nov 12th at 5¾ miles. We then went on until we had made 7¾ miles and built twin cairns at our lunch camp. The surface was very deep and soft. From this camp we can see back to our last horse walls 7¾ miles geogl[5]. There were good iridescent thin filmy cr. st. slips all over and round the sun at lunch in which the colour was greenish blue and pinkish violet chiefly, more or less in rings suggesting a corona more than a halo, but gradually diminishing outwards in intensity so that at no point could one say there was an outer limit to the colours.

This was at 3.30 a.m. app. time. During the second march Nobby and I had a disagreement and he lodged one of his objections on my heel — without hurting me. After this he went exceedingly well. We turned in on a glorious hot morning at 10.30 a.m. Our horse walls arranged thus

$$3+1 \qquad 4 \qquad 2$$

We had passed a cairn of Evans' at 3¾ miles before camping here for the day.

Mon 20-Tues 21 Nov   Turned out about 8 p.m. and got away at 11. Made 7½ miles by lunch time passing a night camp cairn of Evans' at 1½ miles and lunching at another of Evans' cairns which we converted into a twin cairn. All this march the sky has been becoming overcast with radiating cir. str. streaky cloud forming gradually from the S.E. and overcasting us and then extending to our west. The land at the same time has again become brilliant and clear — whereas when we were having clear sky overhead the last two days the land to our west was overcast with dense stratus. The surface today has been a trifle better — patches of hard road-like crust which carries the horses appeared for the first time last a.m. march. This p.m. march they have again been with us. No wind — light var[iable] airs. Sastrugi in the main E. and S.E. After lunch when we made a twin cairn — mentioned above — we had another 5½ miles reaching a cairn of Evans' and just passing it before building our horse walls.

Clouds high cir. st. radiating S.S.W. and N.N.E. and being caught by an easterly high current north of us. Sastrugi here show pretty strong S. Ely wind lately. We are now turned in with gentle S. Wly breeze and sunshine. Temp. +2°. Hot sunny day.

Tues 21-Wed 22 Nov   Turned out about 8 p.m. and got in 7½ miles before lunch, passing a camp of Evans' at 4¾ miles and building our twin cairns at lunch, we then could see Evans' camp and an enormous cairn by it built at 80°31′ which we reached soon after 2½ miles after lunch. Theodolite and 2 boxes biscuits depoted here. We found the motor party, Evans, Day, Lashly, and Hooper all fit and well and very hungry indeed. They had been waiting here six days and had read Pickwick through. Day looks thin — the others not so thin. They came along and camped with us at the end of our 5½ miles, where we built

The cairn Evans built at 80°31′ was about 15 ft high — enormous — and has a flagstaff. Two boxes of biscuits depoted here. Jehu is to go on 3 more days and then beats the distance at which Shackleton's first horse was shot[6]. The motor party continue with us till then, and then Day and Hooper return. The dogs go on with us indefinitely towards the glacier foot. Cloud radiation today is still from S.S.W. to N.N.E., but is clearing away. Surface improving slightly, still very deep when the horses go through crust. Horses all going very solid and steady. Nobby is here given an extra box of biscuits. Sastrugi strongly marked S.S.E. or S.E. Hot day of sun and no wind for our sleep.

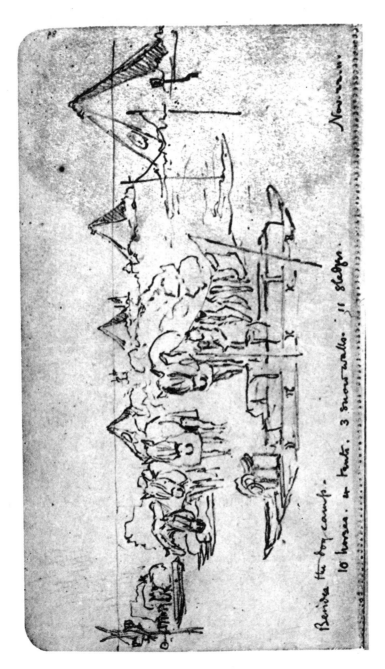

Wed 22-Thurs 23 Nov   After a long confab in our tent, attended by Scott, Cherry, Bowers, Evans, Day and Titus, and Meares and myself, we got away about 10.30 p.m. and did 7½ miles passing a cairn made by the advance piloting and crock pony party at 4½ miles, and then making a twin cairn at lunch. Clear hot sunny night with light following airs from the N. All the ponies going well. Surface rather deep for them but not very bad. The surface crusty with crystals which allow sledges to glide well as soon as the sun gets on them — but at night they are sandy and sticky. In the a.m. march we did 5½ miles passing a cairn at 2¾ miles and making horse walls at the final camp.

Thurs 23-Fri 24 Nov   We made 7½ miles in our p.m. march after a good sunny day's sleep, and the advance guard built a cairn at 4 miles about and a twin cairn at lunch. After lunch we did 5½ m. passing a cairn at 3 miles and our horse walls went up

There was radiating cir. st. forming all night across from due E. to N.W. and cloud banking up thick in the S. with a S.E. puff occasionally and a Sly breeze in the morning when we turned in. We have had some fine iridescence of high cirr. today and yesterday — lemon yellow broad band nearest the sun, with pink and vivid turquoise sort of prussian emerald blue outside, contrasting in its greenness with the French blue of the clear sky and the pure white cirrus. Again today we had grey windy and radiating high cir. st. with a break here and there showing still higher cirrus cloudlets beautifully iridescent with pink centres and green borders (see sketch)[7].

Fri 24-Sat 25 Nov   There was a strong appearance of approaching blizzard weather from the south all night and complete overcast low stratus and a few falling snow flakes when we started off in the evening. Made 7½ miles, passing a halfway cairn and making 2 at lunch. We had good surface and all the ponies went well. Made 5½ m. in the morning march and Titus shot Jehu when we arrived in camp. Jehu is for the dogs. 81°15′ is our position today and the sky has completely cleared. Very hot sun and blazing glare. Light Sly air. During our first march there was at first a S. Ely breeze and then a S. Wly. Our course from 80°35′ has been S. 9 E., or N. 12 E. magnetic. Jehu has been brought along a good few miles further south than the lat. where Shackleton shot his first pony. We have 9 left and all going well, though Jimmie Pigg and Chinaman are slow. Considering that when we left Cape Evans Atch was started off a day ahead to see whether Jehu could walk 15 miles without dying, he hasn't done so badly. We got away and did 7½ miles before lunch, passing a midway cairn and making twin cairn at lunch. After starting on our second march we had some white fog and an attempt at a fog bow — also very fine solar corona in 3 rings, reddish orange, greenish and bluish, then orange greenish and blue again outside that and so on — formed by the low fog — and the strange thing is that the iridescent high cirrus round the sun appear to be the same to some extent colour, and the similar coronae irregularly arranged. There was another phenomenon today which we called the *Garrard halo* — a sparkling ring round the sun, but only a few feet from one's eyes formed amongst the glittering crystals amongst which we were walking with brighter mock suns as usual on each side but with no colour and the discrete crystals all visible. The surface today was very heavy and looked as though a profuse growth of bunches of crystal points and plate points had grown on it. And on the N.E. side of the sastrugi everywhere there was 3 or 4 times the growth and the crystals were arranged in tiers. All the horses went well and Atch joined the man-hauling pilot party. Lashly joined our tent and we are now for the first time 3 units of 4 men each. 1. Scott, Cherry-Garrard, Lashly and myself. 2. Oates, Bowers, P.O. Evans and Crean. 3. Evans, Wright, Atkinson, and Keohane. Meares and Demitri follow us with the dogs. Jehu cut up into 4 days' rations for 20 dogs. Day and Hooper left us this day — no, yesterday evening, Nov. 24th — and turned home taking with them a letter from me to Ory. Turned in now about midday 25th to sleep.

Sun 26 Nov   We start off later tonight by two hours, really at 1 o'clock on Sunday morning, so the marches are both a.m. of the same day now. We made 7½ miles with a midday cairn and at lunch depoted a week's provisions for 3 units in a large central cairn with bamboo and black flag and 2 small cairns one each side, not square with line of march. After lunch it became overcast and snowed and we did 5½ miles making a midway cairn and horse walls as usual. Started my old depot finnesko today with holes in and got them full of snow and water. Surface not so bad. Horses going well all of them and loads lighter now by far. Nobby has now barely 500 lbs = 6 biscuit boxes + 100 lbs of oil + 99 lbs standing weight. Halo 42° today caused by stratus and corona with 2 rings caused by low fog against the stratus. No view at all. Read H.C. and other things in the bag.

Mon 27 Nov   Turned out to our breakfast meal at 1 a.m.
Monday. Snowing more or less and a cold breeze chiefly S.E. Much
windy cirrostratus overhead and later on overcast blind white
weather. Soft surface, but horses went well. Did 7½ miles leaving a
midway cairn and a twin lunch cairn. Capt. Scott took over the
remainder of my week's cooking to save his foot which is painful.
Chinaman is to be shot after tomorrow's march and then Lashly
leaves our tent again and Keohane joins it.
We had very heavy going for the second march. Snowing all the time
could see nothing and the course was very erratic in consequence. We
made a midway cairn and horse walls as usual. Still snowing and
breezing from the S. when we turned in about 4 p.m. Capt. Scott
cooking. We turned out this morning at 3 a.m. Tonight we turned
out at 4 a.m. gradually working into day routine for the glacier
which we should reach in a week and then finish with the ponies.
Snowing pretty well all the time we were camped.

Tues 28 Nov   We turned out at 4 a.m. and got away at 6 a.m.
Marched in a sort of summer blizzard with falling snow and some
drift but no great force of wind — what there was was due S. and
sometimes it blew a fresh breeze. Temp. +12°. Last camp it was
+17°. The snow falling is damp and clogged the ski heavily when
the pilot party tried them yesterday. Course today altered to due S.
i.e. N. 20 E. again, as we are getting bad deep snow surface anyhow
so we shouldn't gain much by keeping out. We can see nothing today
— overcast everywhere and white. We did 7½ miles leaving a
midway cairn and twin lunch cairns. Course has been very erratic
yesterday — afternoon especially — and this first march today not so
erratic. After lunch made 5½ miles and the surface was better, due
to Sly wind, but the march was in falling snow all the time often thick
and warm, melting where it fell on the sledge. Shot Chinaman this
camp and Cherry took on the 10 ft sledge, leaving his Michael's 12 ft
sledge depoted here over the horse remains. Meares and the dogs are
up now almost at same time as we are. Now that we have turned in
the sky has cleared and the wind dropped, sun out and things look
brighter. No land visible, though in the middle of the night Mt.
Markham's 3 peaks and Mt Longstaff appeared fairly clear in places,
but were obscured later and all next day, though we had a good
sunny march 7½ miles in the first half, then lunch and then 5½
miles.

Wed 29 Nov   The surface was deep, wettish snow into which the
horses sank from 8 or 10 inches to a foot very often and we ourselves
had heavy walking. The horses all went well. Wright has now joined

the manhauling party. There were once again continual settling crusts in the snow we were on today. They sometimes made a noise like 'hush' sometimes like ripping canvas and sometimes like thunder crackling and then rumbling — but of course not so loud. The cracks are quite visible rambling in circles round our line of march and the weight of the sledges cause[s] them. They occur generally when recent heavy drifts are stepped on — and today the snow on many parts has been heavy and wet packed.

Thurs 30 Nov   We have today passed our previous furthest south and are now camped in 82°21′S. We could see Cape Wilson from time to time. Glorious sunshine all day but often a deep heavy surface into which the horses sank sometimes to their knees and almost up to their hocks. They are tired after it, especially Christopher, Victor, Snatcher and Michael. It was heavy walking for us too. After lunch we did 5½ miles and a little of the lower land showed, but nothing to sketch. Very hot. Tried Nobby with snowshoes today, but they came off every time. Otherwise they would certainly be very useful in saving him from going in. We are in our bags now by 4 p.m., turn out about 2 a.m. Tomorrow the tents rearrange a bit. Titus comes into ours and Cherry goes into Bowers' tent.

Fri 1 Dec   Turned out at 11.30 p.m. and was sketching the coast which came out beautifully clear until 2 a.m. when we had breakfast and got under way. We are about 40 miles out from Mt. Longstaff and can see Mt. Markham's peaks beyond and Christmas Mt. to the north and Cape Wilson and Capes Goldie and Lyttleton. Longstaff is a magnificent mass of mountain — not much rock showing and what there is of two kinds, the bulk of what shows from summit to base being very dark, almost black, while there was also a pale chocolate buff rock in vertical cliffs exactly as we saw this rock somewhere further N. by Christmas Height in 1902. There was this black rock both above it and below it and the two sides of the glacier coming down steeply from Mt. Longstaff has a 45° scarp on the N. side of the black rock and a similar snow covered scarp on the S. side where only this pale chocolate buff rock shows. This glacier is tremendously broken up like an icefall. This is the more northern of two glaciers coming down from Mt. Longstaff. The other one descends to the south of the mass of reddish cliffs. The whole range is a magnificent mass of mountains. We made 7½ miles in the first march. Nobby on snowshoes for 5½ of them. They broke on the start of the afternoon march, so took all off. We then did 5½ miles and now are turned in.

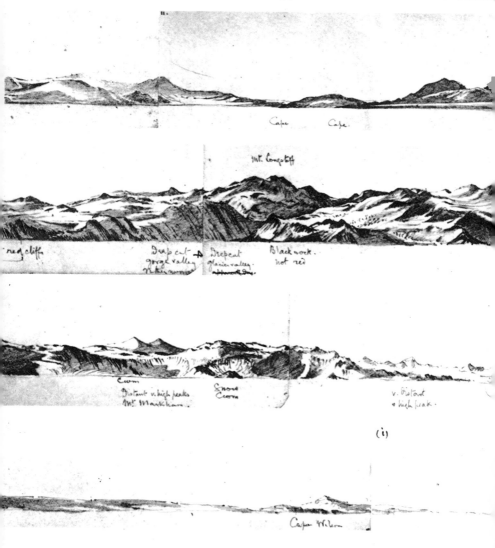

Dec. 1. 11.   2 a.m. finished ?
Nov 30. 11.   11.30 pm. begun. ∫ (a)

farthest land visible
to the left.

Panorama of the landward approaches to the Beardmore Glacier

Cape.   Cape.

Mt. Longstaff

red cliffs   Deep cut   Deep cut   Black rock.
             gorge valley   glacier valley.   not red
             ↑ this morning

Cwm
Distant v. high peaks   Snow   v. Distant
Mt. Markham.   Cwm   + high peak.

(i)

Cape Wilson

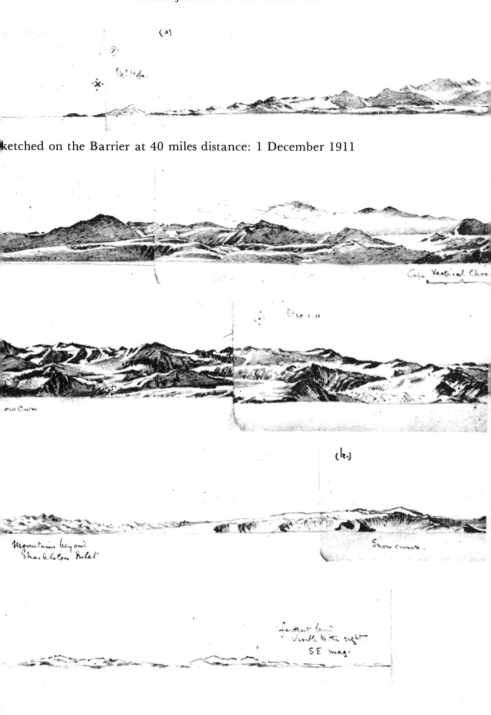

ketched on the Barrier at 40 miles distance: 1 December 1911

Overcast now after a very hot sunny day, the clouds having come from the S.E. All the sastrugi still from S.E. winds. No settlement of the crust today at all. Titus has come into our tent tonight and we have lost Cherry who has gone to Birdie's tent. Christopher shot tonight. We had some of Chinaman in our supper hoosh — and it was good like boiled beef — put in when water was cold in very small chips and boiled up with the pemmican. Here we are depoting a week's man food and a sledge and Christopher's bones.

Sat 2 Dec   Absolutely blind day, no land visible all day. Made 7½ miles first march and 5½ miles in the second and now turned in 7 p.m. In the evening Victor was shot and cut up. We had a big meal of horse meat off Christopher tonight which was very good eating indeed, very tender and very slightly sweetish taste unless plenty of salt added, when it was just like beef to taste and sight. The surface all day was deep soft snow. The horses all started badly. It was very close and hot and snowing but when a light breeze which was Nly became Wly they went very well all the rest of the day. Still snowing and overcast and blind glare on turning in. No land visible. We are crossing undulations now without a doubt.
The snow is deeper and softer in the hollows than on the summit. I should say we crossed from one valley into the next in half a mile or less.

Sun 3 Dec   Advent Sunday. We awoke to a blizzard blowing from S.E. or S.S.E. Thick drift and heavy snowfall. We therefore lay in till noon when the sky cleared and we got off a march after lunch. We then saw Mt. Hope and the Gap[8] quite plainly but an hour after starting everything became obscured by low strat. from the S.E. again and we walked 10 miles in deep, soft snow by 7 p.m. when we camped without having seen a shadow or any land at all. The course is therefore erratic. Scott and Bowers having no horses now went ahead on ski. Evans and the manhaulers camped at 6 miles. The dogs came up to us at 10 miles. Still blowing and snowing hard from N.W. when we turned in after a hoosh with plenty of Victor in it. The horses are going splendidly still, though their walking is very deep. The sledges pull light enough. Sastrugi not much marked, but all S.E. and today's N.W. wind is under cutting them into tongue shapes. Heavy banks of stratus in the S.E. here about seems the invariable sign of bad weather.

Mon 4 Dec   Blew all night and at 3 a.m. after about a quarter of an hour's lull the wind went round from the N.W. to the S.E. and by 7 a.m. we were having a regular thick blizzard. We lay in till 2 p.m.

having built a new wall for the horses on the S.E. side. Heavy drifts of soft snow. We got away at 2 p.m. when the sky suddenly cleared and we could see the land ahead to the S., Mt. Hope and the glacier and all the land beautifully. There were several magnificent valleys with glacier icefalls coming down. We marched solidly on till after 8 p.m. when we had made 11 miles, then camped below a very big undulation 12 miles from Mt. Hope. We had the long line of pressure caused by the Beardmore Glacier in view on our port bow all the time and we crossed 3 or 4 big waves of undulations in the last two miles; they were very marked. While on our right we could see the dark lines of chasm splits toward the coast. This evening the wind seems again inclined to get up from the S. and there are marked radiating clouds from the S.S.E. and N.N.W. beginning with high spidery cirrus, and now much lower with squally wind and surface drift. Michael was shot this evening. We are still eating Victor. One of the peaks on our starboard bow has bands of horizontally bedded rock at the top like Lister. There are fine craggy cliffs of a dark red rock — looks like granite to me but is said to be like Cathedral rocks which are of dolerite. Three glaciers come down 3 valleys together on our starboard bow, all heavily crevassed and icefalls — no moraines showing anywhere but a heavy curved line of pressure always on the inner side. On our right beam is a part of the foothills where the ice appears to have first carved out scarps and ridges and later to have sheared the top of the ridges off into a plateau form.

It was midnight nearly when the first chance came to draw, and then not only was the sun due S. making everything in shadow and formless, but the wind drift got up.

Tues 5 Dec   By the morning we could see nothing. We were in a thick blizzard blowing very hard at times and snowing very heavily, so that we had to lie up the whole day. It was +31 and very wet, indeed sopping in the tent. We turned out periodically to feed and dig out the horses who were nearly buried in drift. The wind is all from the S. and S.S.E. and the blizzard was evidently coming down the glacier. Last night there were drift clouds off the tops and a white fog outlining all the upper heights making their outlines hazy, especially the rocky ones. Our sledges have been completely buried out of sight in some cases as we are in a depression.

Wed 6 Dec   Blizzarding like blazes all night and again all day making it impossible to proceed. The horses are feeling it now — they are wet through, poor beasts, but they feed all right. Excessive drift, we have to dig them out every 4 or 5 hours. Our tent is more sopping than ever and our windproofs are soaked with water every time we go out as the temp. is $+33°$ and the snow like heavy wet sleet. Bags getting pretty wet in consequence, so is all our gear. I sleep most of my time lying up between the meals and seeing to the horses. Also have read a good deal of Tennyson — *The Princess, Maud*, etc. — in a volume I brought. Wind still from S. and S.E. Very wet and heavy snow.

Thurs 7 Dec Same weather all night though it lifted late yesterday evening enough for us to see some of the land. It still blew and snowed and this continued all night and all day today — no change in the wind and heavy soft snow driving all the while. Our wind clothes are absolutely soaking wet and we only put them on to go out, and take them off at once on turning in to the tent again. We have all begun our summit rations today — (a new box of biscuits and a new can of oil). We didn't open one biscuit box until for the supper meal of the 7th today, but it was to begin at lunch when we ate up our wet scraps. We filled our primus yesterday after breakfast, and still have a meal in it and a whole fill in the can which goes to Meares. This can was used up during the blizzard. We didn't open the new one on the 8th and breakfast of the 9th came out of the *old* one.

Fri 8 Dec We woke up to the same blizzard blowing from the S. and S.E. with warm wet snow $+33°$. All these three days frightfully deep and wet, and tremendous great drifts which have completely hidden several of the sledges and topped the pony walls. We turned out after breakfast but after digging out the sledges and all our gear, trying one of the horses on the deep surface and finding he went in up to his belly, and after trying weights on a 4 man sledge with and without ski, we decided to wait one more day for the weather. We then shifted all our tents and made a new camp to windward, had lunch and lay in all the afternoon while the weather very gradually improved — first the wind dropped and we had variable airs and gradually the snowfall began to break off and on, though the snow continued to fall all day and is still falling now from the N. and N.W. with a light breeze.
Everything soaking wet, sleeping bags sodden, tent dripping, water everywhere till late this evening the temp. dropped to $+29.7°$ and the snow began to diminish. It has been oppressively hot and steamy all day, making one sweat digging things out in a few minutes, have been living in pants only — windproofs all wringing wet, so also

sleeping boots, socks, gloves, everything. Tobacco juice running brown out of Titus' private gear bag. It has been a phenomenal warm wet blizzard different to, and longer than, any I have seen before with excessive snowfall. Knee deep away from the camp and Nobby's belly touching the surface every step, poor beast. The 5 horses still living are on a very low ration, but it is mainly oilcake. I have kept Nobby my biscuits tonight as he is to try and do a march tomorrow and then happily he will be shot — and all of them as their food is quite done. We are just 12 miles from Mount Hope. Have been reading and re-reading Tennyson's *In Memoriam* this blizzard and have been realising what a perfect piece of faith and hope and religion it is, makes me feel that if the end comes to me here or hereabout there will be no great time for Ory to sorrow. All will be as it is meant to be, and Ory's faith and hope and trust will be to her what Tennyson's was to him. But *In Memoriam* is difficult reading, and the beauty of it wants pains to find, but it is splendid when found.

Sat 9 Dec   A very long tiring and bad day though we got on the move at last and made 8 miles on the worst possible surface. We first of all started off the manhauling party to make a track in the deep soft snow for the ponies and then we tried to get the ponies on in this track with their loads. It was horrible work flogging them on, floundering belly deep as they were stiff from the 3 days' blizzard on low rations, very hungry and with a mere apology for a ration this morning, and no more food for the end of this march only a bullet for each. We marched solidly on with great difficulty and great exertion from 8 a.m. till 7 p.m. No stop and no food for ourselves, or the ponies. Nobby had all my 5 biscuits last night and this morning and by the time we camped I was just ravenously hungry. It was a close cloudy day with no air, and we were ploughing along knee deep alternately hammering and dragging and encouraging our poor beasts — for we couldn't have moved these loads ourselves on this surface. It was beastly work and the horses constantly collapsed and lay down and sank down, and eventually we could only get them on for 5 or 6 yards at a time — they were clean done. Then we camped. Shot them all and then before turning in had an hour or two of butcher's work cutting them up and skinning them for dog food and for a depot for ourselves for our return journey. Thank God the horses are now all done with and we begin the heavier work ourselves. We left 4 sledges here, 3 10 foots and 1 12 foot on end, with some personal gear under the left most, as one goes home, and some meat under the next I believe. I left here my pyjama trowsers and my spare pants, some socks and mits and ski straps. Slept like a log in my still very wet bag and woke fit as a fiddle to a perfectly glorious hot morning.

Sun 10 Dec   Blazing hot and only slight draughts now and then amongst the mountains. All yesterday we were approaching the Gap entrance gate to the Beardmore Glacier and Mount Hope on the left with Mt. Asquith and Cape Maude and Cape Allen on our right. Magnificent ochreous reddish gneissic granite columnar crags and pillared mountains on both sides, the western end of Mt. Hope being low and rugged with the same nearly all in situ — not much scree anywhere and what angle the valley sides make is always about 45°. Enormous cwms on our right low down and full of snow with avalanches here and there and in one or two places the screes ended at the top in a sort of sheared off table-land of rock. The mountains of gneissic granite look almost like columnar basalt. They are weathering out vertically and their tops are flat at about 3,000 ft with a dip to the west about so much.

Mt. Hope is much rounded by glacier action and so is the smaller rocky foothill on the W. side of the glacier. However this is all yesterday's observation. About the pressure in the barrier ice — it is immense where the Beardmore exits east of Mt Hope. Then comes our narrow entrance W. and then more pressure along the Maude Allen coast. We crossed some ridges and hollows all deeply snow covered and found only 3 or 4 crevasses, but Snippetts, one of the horses, very nearly disappeared in one of them — got all his hind quarters down. We unhitched the sledges and then flogged him out of it, and he scrambled out — but the crevasse would have taken any number. I made a circuit round and though Nobby must have crossed the same crevasse he didn't go through. This of course was on Dec. 9th, just before we shot them all.

Now for the 10th. Blazing hot. We made the top of the Gap slope partly on ski and partly on our feet — very heavy hauling. 600 lbs per sledge, 4 men each — 3 sledges — very soft deep snow — and then lunched. Here Atch told me that Silas and Lashly are knocking up with the heavy work they have been doing. We ski-pulled all the afternoon and made 6 miles altogether and camped in a snow hollow where we got caught in a very brisk drift storm — much dry soft powdery snow flying all round everywhere. We shall make a depot

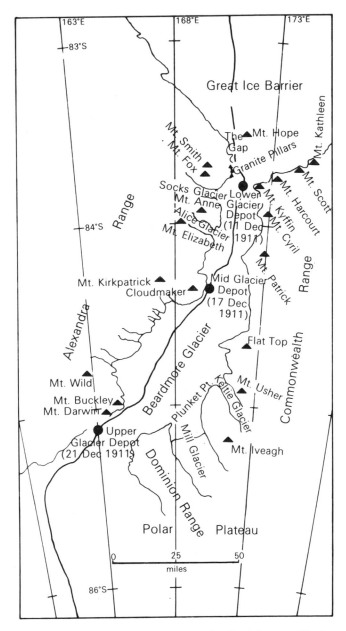

Map of the Pole Party's route up the Beardmore Glacier

131

here of return food and go ahead tomorrow still taking the dogs for half a day further. We have come through the Gap now, and on both sides have had magnificent ochreous reddish columnar crags of gneissic granite weathering out like basalt vertically in columns. We are now camped with a view right up to the inland ice and down the glacier pressure to the Barrier, a fierce sight of pressure but wonderful and magnificent. As Meares returns tomorrow I must send a note to Ory by him.

Mon 11 Dec   We made our Lower Glacier Depot in fine sunshine and raised a cairn and a flag on it, black bunting. Meares came on with us for the forenoon's march and we on ski in 3 parties made for the middle of the glacier in a line mainly for Mt. Kyffin. We crossed some waves of pressure where the shiny blue ice showed up like combs here and there and P.O. Evans just behind me got into a crevasse with the whole of one leg and ski, the whole length of which broke through. We had 3 or 4 miles of this — not bad at all — and then the surface flattened out with very occasional irregularities only. We lunched and Meares and Demitri then started for home with my note. Their dogs all very fit indeed. They ought to have a very easy journey. I sketched in the morning and after lunch and on the forenoon march visited a large boulder isolated in the pressure ridges and found it to be very coarse granite full of large quartz crystals an inch and a half square and white quartz veins, very full of mica and hornblende and quartz, and some isolated sort of inclusions of fine grain grey gneiss. We had full loads in the afternoon when Meares left us, 680 lbs or more. Scott tells me we are pulling close on 200 lbs each. We managed this better and better as we went on in the afternoon and eventually camped in very deep soft snow — up to our knees abeam of Pillar Rock. We had to dig down nearly 3 feet for hard snow for the cooker. Birdie's party came up soon after us. Teddie Evans' were still labouring up when we had our bags out, so Scott and I went and helped them up very late. Birdie, Teddie Evans, Keohane, Lashly and Titus Oates have all got pretty bad snow-glare. Both sides of the glacier at present all same reddish ochre rock — probably granite — in vertical column with bands of quartz. Kyffin side the same evidently, but below Kyffin is some blacker rock. Notice almost complete absence of screes of loose rock — all in situ. Also notice banded rock like Beacon Sandstone[9] top of high mountains to W.S.W. Also notice dip of flat top of red ochre rocks on our right to be so

132

We had a very heavy forenoon's pulling on ski — the sledge constantly refusing to budge at all.

Tues 12 Dec   We turned out about 8 a.m. No crevasses, no ice — nothing but soft and softer snow with an occasional mound where irregularities lay underneath. We were going at first a bit towards Kyffin, but later towards Cloudmaker and then even more to the right. See sketch made at lunch for detail of left side of our view. After lunch we got on much better — same soft surface though, and camped at 7, in snow up to our knees as we walked. Couldn't have gone a mile in this on foot hauling without ski. We are now abeam of? Socks Glacier, a basin outfall of ice on our right, while on our left we see the great disturbed ridge shot out just S. of Mt. Kyffin. Between us and it a huge névé hollow of soft smooth snow which we are avoiding. We can today see the Dominion Range at the very head of the glacier just above the horizon. We have passed the Pillar Rock and red cliffs. Both parties got on fairly well today, but the work was very heavy indeed. Now turned in 9 p.m.

Wed 13 Dec   Turned out 6 a.m. and had bad light to begin with but broiling sunshine by lunch when we had done about 2 miles on the most awful surface. At lunch we fixed on the spare 10 ft runner and tried in the afternoon, but though the runners were a help the surface had got sticky and we did almost nothing — about a mile — the surface alternately excessively soft, and soft enough to just hold the sledge up one side only. It was killing work when the sledge hung up as it did every 20 to 30 yards. The other 2 sledges did relaying for the last bit — the first time. I had a grand 2 or 3 hours' sketching at the lunch camp, and did Mt. Elizabeth with Socks Glacier and Mt. Kyffin. Turned in a quarter to ten.

Thurs 14 Dec   Turned out 5.30 a.m. and got in a real good day's travelling still on deep, soft snow, but the sledges went on it, all 3 of them. We still have our 10 ft runners under the sledge and we exchanged sledges with Birdie's team to see whether they found ours heavier or lighter. We both thought ours was the heavier pulling, though the weights are identical, yet our team, Scott, Oates, P.O. Evans and myself — walked down both the other teams every time. The other teams are Teddie Evans, Atkinson, Wright and Lashly — and Bowers, Cherry, Crean and Keohane. We constantly struck blue hard ice with the points of our ski sticks covered by a few inches only of soft snow. The surface has improved a lot — much firmer. We rose 570 ft during the day now being 2,000 ft above the Barrier and

Dec. 13.11. 3 pm. Lunch Camp. No. 1.

Mt Elizabeth. 10761 ft.

Rock shews hardly anywhere but on the crests. 4 tops

Dec. 13.11. 3 pm. No. to 10.30 pm Lunch camp

Heavily Crevassed

Very heavy Ice Fall

Dyke Coulridge.

Socks glacier.

at about Shackleton's position of 8½ [10]. We have passed Kyffin and Socks Glacier and are opening up more banded mountains ahead. The Cloudmaker shows its very carved out profile, hard and softer rock — very wet with a very hot march, now overcast. Northerly breeze. Turned in at 9 p.m. all satisfied. We began 8 biscuits a day today. Camped on soft snow but only a foot or so of it above hard blue ice and several crevasses — small ones stepped into camping. Got a sketch made at lunch time.

Fri 15 Dec   We had a good day's march — ski and all three sledges running well — but gradually all land got overcast and we had to camp an hour before time as snow fell and hid everything for steering by. We passed Alice Glacier (see notes)[11] which I wish I could have sketched, but all became obscured. Surface getting much harder — windier sastrugi and only a few inches to less than a foot of snow above the blue ice everywhere all day. Lips very cracked by sun — bleeding and sore — nose and face also blistered and scabby. Most of us are the same — but very fit and enjoying it. Have passed Alice Glacier opening now.

Sat 16 Dec   We got away at 7 a.m. and though it had been overcast and snowing and a N. Ely breeze the sky was clearing when we started. We got in a good 5 miles before lunch on ski, but after lunch there was a cold S. Ely breeze and the surface got so bad for ski that we took to hauling on foot — the snow being very trying only not deep above the blue ice — there was a very trying crust which gave way when you hauled on it. We found lots of crevasses with one leg, but nothing worse, and by 6.30 p.m. when we camped we were fairly well up to the great ridges of crevassed pressure caused by the Keltie Glacier running into the main glacier and in sight of the moraine running down from the Cloudmaker. The Cloudmaker has bands of black rock running across it. Got some sketches at lunch time. Now turned in 9 p.m. We are camped right up against bare ice pressure ridges.

Sun 17 Dec   We made a good march over a long series of high pressure ridges to the westward, all the forenoon tobogganing, then on crampons all the afternoon, on a hard surface with an infinite number of cracks and fissures into which we put our legs and feet. I had a rotten afternoon with snow glare, streaming eyes and at times nearly blind, but got through all right and had zinc sulphate in both eyes that night and as a consequence was awake for about 6 hours with pain. Got 2 or 3 hours in the early morning. Depot laid.

Mon 18 Dec   We again had a fine hard rubbly ice surface with any number of fissures and cracks to step in, all bridged ones but many rotten. It was overcast and snow crystals falling. My eyes better but couldn't stand any light without green and brown goggles.

Tues 19 Dec   We got on the same surface again to begin with and later in the forenoon on to hard névé mixed with smooth blue ice and sastrugi; a transition from glacier ice to the summit surface. The rubbly blue wavelets in the blue ice lower down are here formed in opaque white ice. The sastrugi of size are all made by S.W. winds. We are now nearly up to Shackleton's 16th position and have Mt. Buckley ahead of us on the island which stands between Plunket Point on our left and Mt. [   ]¹² on the right. The whole range on our right is magnificent series of stratified rocks terraced and carved by glacier, black and broad bands of paler yellowish rock. As we keep as far as possible in the middle of the glacier we get no chance of seeing rock or moraine. We made a long march on ice nearly all the way up 800 ft and covering over 20 statute miles.

Wed 20 Dec   Starting 8 a.m. to 1 p.m. and from 3 p.m. to 7 p.m. we then camped on 85° lat. — 300 miles from the Pole just below a rise steep up to the main pressure here east of Mt. Buckley. Fine cliffs of banded rock. Crevasses of small size abounded all the day but not more than 6 in. to 6 ft across, running mainly S.E. to N.W. and at right angles to these old silt bands which had been jammed out of line and were very much older than the crevasses. Much of the ice was marble smooth. Birdie and I at lunch hour walked about 3 miles to look for the sledge meter parts on the back track, but couldn't find it. We had fog rolling up from the N. at lunch, but it cleared again during afternoon when wind dropped. Rock stratification magnificent everywhere, but little chance of doing anything at it. Now abeam of Mt. Buckley and turned in 10 p.m. The first supporting party returns tomorrow night — Wright, Atkinson, Cherry and Keohane.

Thurs 21 Dec   We had a very fine long hot day's march over much blue rugged ice and crevasses everywhere, in fact a regular day out, and as has happened the last several days we had a Nly breeze before noon which brought up a thick white fog and made it impossible to go on. We camped for lunch in a maze of crevasses and waited an hour or two till it cleared and then got in a very good finish to our march and reached a good place for the Upper Glacier Depot where we part with the first returning four. A flag was put up on the cairn of snow. All I depoted here was my sketch book — in an old pair of finnesko — and my crampons. Also a bottle of brandy to be picked up with a medicine chest by the second returning party. It was wretched parting with the others. Atch took my watch back as they were short and had only one. Silas took my sundial. Atch also took a letter from me for Ory.

Fri 22 Dec   We made a very good march — the two parties 4 each — in good weather and on a good surface, nearly all up hill and in a S. Wly direction with pressure ridges on our left beam and left bow all day. And the Dominion Range almost behind us on our port quarter. Sketched in the evening.

Sat 23 Dec   We made a large and a small cairn and started away at 7.45 a.m. first down a flat, then up a slope over a ridge across another flat and up another ridge, and so on I think we crossed about 5 of these ridges in all, today 3 and yesterday two. Camp tonight 7,700 ft above sea level. These ridges abut on the two pressure ridges formed by the movement of the summit ice down the glacier. We made over 8 m in the 5 hours' forenoon march. Temp. about −5° and a hot sun with cool Sly breeze all day. We made 7½ m more in the afternoon in 4 hours and are camped now on what certainly looks like the summit. Endless flat plains of snow lie before us to the S.S.W. and W. and just the tops of disappearing mountains behind us. On the ridges we have had a lot of nasty treacherous crevasses down which all of us put our legs in turn. Twice we had greenhouse ice[13] with a false bottom — very disagreeable to go over. We have also crossed many wide crevasses bridged well, but sunk and with very rotten lower edges this time, instead of upper edges as we had on the glacier below.

Sun 24 Dec   Christmas Eve. Fine and sunny with a good strong Sly or S.S. Ely breeze in our faces all day. Temp. −3°. We got in 14 miles in 9 hours marching and rose nearly 400 ft passing a few little ridges of pressure, but not a single crevasse and going due S. all day. Very promising — thoroughly enjoyed the afternoon march. Surface on top of the rises very white, smooth marble ice, otherwise flat crisp firm snow taking impressions well, but only a half to one inch deep, interspersed with pretty big hard sastrugi from S.S.E. Can still tonight see tops of one or two mounts especially Flat Top — very high to the N.E. of us now.

Mon 25 Dec Christmas Day and a real good and happy one with a very long march 17½ statute miles over two or 3 rises with any number of crevasses and greenhouse surface. Lashly had a straight drop the length of his harness. It was blowing 3 — 4 all day in our face — S.S.E. with surface drift — but otherwise was very bright and sunny with a clear blue sky which we get every day up here. Temp. — 7. We have not risen much today, 250 ft, but in passing over one of the rises we were apparently passing over a mountain top and the highest point had a great bump of ice pressure with an enormous hole on the southern side. We had a magnificent lunch of 3 biscuits, 1½ pannikins of tea, a spoonful of raisins, a whack of butter, and a stick of chocolate. For supper we had a regular tightener. Started with pemmican and horse meat and onion powder and curry powder and biscuit dust hoosh — 1¼ pannikins. Then a pannikin of arrowroot, cocoa, sugar, biscuit dust and raisins, then a pannikin of good cocoa, then a large bit of plum pudding each, and then 5 caramels and 5 pieces of ginger and 1 biscuit each. We are now all in our bags — 2 hoops[14]. Read H.C. and Christmas reading in bag.

Tues 26 Dec Made 13 geographical miles on a steady going surface and rose 250 ft. We are now over 8,000 ft. We made 7½ miles in the forenoon and the rest in the afternoon. We had the same old south-southeasterly wind in our faces all day today with temperature — 3° in the day and — 7° to — 10° in the evening, bright sunshine — quite pleasant. Worked off our Christmas feast all right. Still see patches of crevassed bumps here and there and still rising new horizons, but got on no crevasses today only on to sort of white marble ice with temperature cracks. No big sastrugi excepting such as are nearly buried in drifted snow.

Wed 27 Dec We had another fine sunny day and good going except for one long stretch of crevasses on a rise where there was a lot of pressure and another great hole at the highest point. The crevasses were badly bridged in places and Titus and the Owner and I were all in at one time. We had less wind and it was hot in the bags in the night. I had a bad short go of snow-glare in the right eye all the afternoon while we were crossing the crevasses, but sleep cured it. Rose 300 ft Temp. — 6.

Thurs 28 Dec A long day. Camped at 7 p.m. Chopping and changing sledges — and pullers — all forenoon to find out what were causes of slowness of second sledge. After lunch we came on with the other sledge and found it infernally heavy. Did 8 miles before lunch

139

and a bare 5 miles after lunch. Fine sunny weather, blue sky. Rose 100 ft only in the day. The other party found our sledge very easy pulling.

Fri 29 Dec   We had a long day pulling at times over a very heavy uphill sandy surface, twice going up a rise where the snow was all like sand. The surface all smooth more or less. No large sastrugi, and everywhere letting one in enough to spoil one's pull. The other sledge changed their loading and came on much more easily. We rose about 200 ft and passed an extensive valley on our right. Sky was well marked radiating cirrus all day E. to W. — first time really marked. Temp. —6° and cold Sly wind all day.

Sat 30 Dec   We had a good day but fairly heavy going. The other sledge fell back a very long way and came in in the evening three quarters of an hour later than we did. We had a short rise and bad surface to begin with in the morning and again a rise and bad soft surface late in the afternoon. Had a long talk with the other tent which is not at all satisfactory.

Sun 31 Dec   We marched from about 8 a.m. to 1.30. The other sledge having depoted their ski at the last camp going ahead of us. We made about 7 miles and came up a very long rise all the way and then camped for lunch near the top and made a cairn with black flag on runner and dismantled the two twelve foot sledges and made up the two ten foot sledges. Evans, Crean and Lashly worked in one tent with the primus going. Scott, Evans, Bowers, Oates and I sat and did various mending jobs in the other tent. We bent the inner lining today for the first time. We had an extra pot of tea between lunch at 2 p.m. and our supper hoosh which was late — after 10 p.m. — as we waited till the 2 sledges were finished. We leave this depot tomorrow with 130 lbs per man without ski. Our unit still has ski which amount in 4 pairs with sticks and shoes to 70 odd pounds.

Mon 1 Jan   A fine day — we had our ski with the new 10 ft sledges and found they worked well — the load came easily. The other party have unfortunately depoted theirs. We are today in 87°6'. Temp. —14°, but feel warmer than usual as there is little wind. We had difficulty in getting into our ski shoes, they were so flat and stiff and frozen. At lunch thawed them out. I got both soles blistered, but we had a pleasant day. Started running the [illegible] from this depot of last night's camp.
We had only 6 hours sleep last night by a mistake, but I had mine

solid in one piece actually waking in exactly the same position as I fell
asleep in 6 hours before — never moved. Tonight being New Year's
Day we had a piece of chocolate each. We have risen today about 150
ft. We are now 9,500 feet above the Barrier.

Tues 2 Jan   A splendid day, −17° last night −11° tonight, but
no wind and the sky is now overcast from the S.E., the first time we
have had any cloud worth mentioning. We were on ski all day —
good surface and the other party on foot. We are still rising but very
gradually. Surface surprisingly flat — very little evidence of high
winds, at any rate lately. We are now about 87°20'. We were
surprised today by seeing a Skua gull flying over us evidently hungry
but not weak. Its droppings however were clear mucus, nothing in
them at all. It appeared in the afternoon and disappeared again
about ½ hour after.

Jan. 2. 12.	Obs.	87° 20' 8" S.	
		160° 40' 58" E.	
			Var. 180°
	DR.	87° 19.8' S.	
		160° 20' E.	

Wed 3 Jan   Blowing pretty hard from S.E. when we turned out
and all day on our march and pretty cold it was. We were on ski all
day, the other party on foot. Last night Scott told us what the plans
were for the South Pole. Scott, Oates, Bowers, Petty Officer Evans
and I are to go to the Pole. Teddie Evans is to return from here
tomorrow with Crean and Lashly. Scott finished his week's cooking
tonight. I begin mine tomorrow. We have come up a hundred feet or
so today. Surface very heavy in sandy drifts owing to the drift
running all day. Looks windy but sky has few clouds. All fit and well.
Writing last letter to go back to Ory tonight before we get back
ourselves. Position tonight should be 87°32' about 148 miles from
the Pole.

Thurs 4 Jan   We turned out as usual at 5.45 a.m. and there was
some delay in getting off as Teddie Evans and his party of Crean and
Lashly came along with us for a mile before turning back to go
home. I was very sorry for Teddie Evans as he has spent 2½ years in
working for a place on this polar journey. We are now 5 and as we
have only 4 pair of ski, Bowers has to go on foot just behind Scott and
myself.
Our sledge is a pretty high pack. We had a perfect day without

141

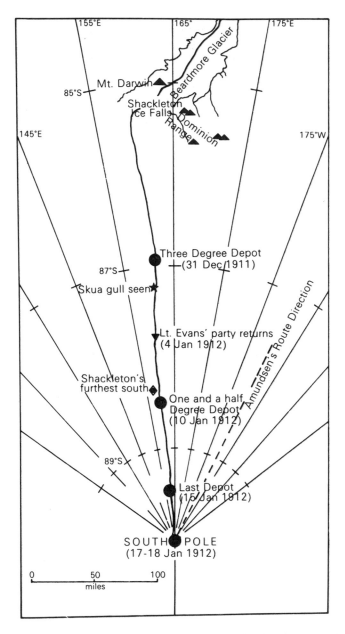

Map of the Pole Party's route from the top of the Beardmore Glacier

wind — calm and hot sun, but temp. —16.8° though so warm I took a bit of sun bath in the evening. The surface was bad, heavy sandy drifts and the sledge went heavy all the afternoon.

Fri 5 Jan　A fine calm day — light Nly and N. Wly airs all day with light but plenty of clouds passing over us from N.W. Sun always visible. Sometimes crystals flying in the air and an effort at one of the brightly coloured parhelia in the forenoon. Pink and green iridescence in the clouds was frequent around the sun. Temperature —14.8°. No apparent rise. Surface heavy all day and sastrugi getting decidedly heavier and more abundant all from S.S.E. and Sly. We had 9 hours heavy pulling on ski — but it was very enjoyable with the following breeze. We sweat freely on these occasions. We ought to be at the Pole on the 15th.

Sat 6 Jan　We had very heavy pulling over the worst wind-cut sastrugi I have seen. All S.S.E. or S. by E., rather and all covered with a growth of bunches of crystals exactly like gorse and best formed on the dagger-like ends of undercut sastrugi. Temp. —22. Last night —23.1, but no wind at all so quite hot in the sun. We began the day on ski, but mid-forenoon took to foot and continued so all day. Tomorrow we leave our ski here as the surface is too much cut up for them and we think it continues so. We believe we came to the true summit late yesterday when we entered the wind cut area. Ice blink all round N., S.E. and W. We lost an hour today going back for a sleeping bag which dropped off, but we made 10 miles and a bit in 8 hours, very heavy work. We are now in 88° 7′, 113 miles from the Pole.

Sun 7 Jan　We were camped in a perfect sea of big sastrugi and so decided to leave our ski here as it looked like the same to the S., but after going a mile and a bit we got out of them, so we stopped and went back and fetched them and continued all day on ski. We thus lost an hour and only made 9.1 miles in the day. Surface very heavy but skiable. Very cold wind met us all the afternoon. Temp. —23. Last night was —27. We get our hairy faces and mouths dreadfully iced up on the march and often one's hands very cold indeed holding ski sticks. Evans who cut his knuckle some days ago at the last depot — a week ago — has a lot of pus in it tonight.

Mon 8 Jan　When we turned out a sort of thick mild blizzard was blowing from the S. or S. by E., with temp. —16, wind force from 4—6. We lay in all day as steering would have been very difficult and the going most uncomfortable.

Tues 9 Jan   We continued in our blizzard camp until noon by which time the wind had gone round to the east or E. by S. with temp. risen to −7 and later in the day to −3. We started off on ski after lunch and made 6.5 miles by 7.30 p.m. on a very good surface with very high deep cut sastrugi occasionally cropping up — all S.S.E. in direction. The whole sky today and yesterday has been overcast but never so as quite to obscure the sun. At this morning's breakfast a sad discrepancy of 26 minutes was discovered between the only two watches. Question is which has gone wrong.

Wed 10 Jan   My last day of cooking. We are now 95 miles from the Pole. We had a very heavy and hot forenoon march and only made 5 miles. The surface was covered with snow crystals which had fallen in the night and the sun on these made the surface heavy as lead. At lunch time we made a depot of provisions with a cairn and red flag. I left my pyjama jacket there.

We had rather better going in the afternoon for 2 or 3 hours, then crystals began to fall again and the sun came out. Good halo and parhelion. We lightened our load at this depot by about 100 lbs. We are now about 85 miles from the Pole.

Thurs 11 Jan   We had a very heavy surface all day — worse in the forenoon — but fine weather, misty to the horizon all round S. and E. and W. with glittering showers of ice crystals, but the sun out most of the day. Radiating cirrus from the S.E. to the zenith and in the evening windy, streaky cirrus in every direction above us, all thin and filmy and scrappy, but a generalized trend E.S.E. to N.W. We had well marked halos of 22° rad. several times but no elaboration. Our latitude tonight is 88° 46′. Temp. min. last night −14°, now −18°. Nothing but light airs today, some of them due south. We make a double cairn each night camp and a single one each lunch camp. For the 5 of us the evening meal takes 2½ beakers of pemmican, 3 spoonfuls of onion powder and 5 pounded biscuits. This makes the hoosh. After that we have cocoa. For breakfast we have 2½ beakers of pemmican and 3 biscuits and tea. For lunch butter and 3 biscuits and tea. The only extras we carry are curry powder, salt and pepper.

Fri 12 Jan   A poor surface but a very enjoyable day's march. Wly and S. Wly breeze came up in the afternoon with overcast of strat. cum[15] made it very cold in the evening, but temp. only −17.2°. Surface was very free of sastrugi — all deep soft crystalline non-coherent snow with no crusts. Min [temp.] last night −26. Horizon clouds all being wafted to and fro a very characteristic effect hereabout.

Sat 13 Jan   Slight breeze all day from the S. and Sly E. with sunshine and intermittent snow crystals. Mouths get dreadfully iced up on the march. We did nine hours getting in amongst sastrugi again early in the day and remaining amongst them all day. Sky blue with white thin wind-blown cirrus overhead and a white low mist of snow crystals all round the horizon. Sastrugi very mixed, the deepest cut being S.S.E. all the while. Temp. −22°. We are now in S. lat 89°8′ − 52 miles from the Pole (C 56[16]).

Sun 14 Jan   A very cold grey thick day with a persistent breeze from the S.S.E. which we all felt considerably, but temp. was only −18 at lunch and −15 in the evening. Now just over 40 miles from the Pole in 89° 20′ about. The surface was not much marked by sastrugi anywhere today and was uniform deep crystalline snow without crusts (C 66).

Mon 15 Jan   Heavy surface all forenoon, but the thick weather of yesterday and the wind are both gone and the sun is out and warm again with temp. down to −25°. We made 6.1 miles in the forenoon and 6.2 miles in the afternoon — 9 hours going. Calm all day except light Wly air in late afternoon and evening. Not a cloud in the sky. Sun about 22° high night and day. We are now 89° 32′ about 28 miles from the Pole. We made a depot of provisions at lunch time and went on for our last lap with 9 days' provisions.

We went much more easily in the afternoon, and on till 7.30 p.m. The surface was a funny mixture of smooth snow and sudden patches of sastrugi, and we occasionally appear to be on a very gradual down gradient and on a slope down from W. to east.

Tues 16 Jan   We got away at 8 a.m. and made 7.5 miles by 1.15. Lunched and then in 5.3 miles came on a black flag and the

Norwegians' sledge, ski and dog tracks running about N.E. and S.W. both ways. The flag was of black bunting tied with string to a fore-and-after which had evidently been taken off a finished-up sledge. The age of the tracks was hard to guess — but probably a couple of weeks, or three or more. The flag was fairly well frayed at the edges. We camped here and examined the tracks and discussed things. The surface was fairly good in the forenoon — 23 temp. and all the afternoon we were coming down hill with again a rise to the W. and a fall and a scoop to the east where the Norwegians came up evidently by another glacier. A good parhelion and plenty W. wind. All today sastrugi have been westerly.

Wed 17 Jan   We camped on the Pole itself at 6.30 p.m. this evening. In the morning we were up at 5 a.m. and got away on Amundsen's tracks going S.S.W. for 3 hours, passing two small snow cairns and then finding his tracks too much snowed up to follow we made our own bee line for the Pole, camped for lunch at 12.30 and off again from 3 to 6.30 p.m. It blew force 4 — 6 all day in our teeth with temp. —22°, the coldest march I ever remember. It was difficult to keep one's hands from freezing in double woollen and fur mits. Oates, Evans, and Bowers all have pretty severe frost-bitten noses and cheeks, and we had to camp early for lunch on account of Evans' hands. It was a very bitter day. Sun was out now and again — observations taken at lunch and before and after supper and at night at 7 p.m. and at 2 a.m. by our time. The weather was not clear, the air was full of crystals driving towards us as we came south making the horizon grey and thick and hazy. We could see no sign of cairn or flag and from Amundsen's direction of tracks this morning he has probably hit a point about 3 miles off. We hope for clear weather tomorrow, but in any case are all agreed that he can claim prior right to the Pole itself. He has beaten us in so far as he made a race of it. We have done what we came for all the same and as our programme was made out. From his tracks we think there were only 2 men on ski with plenty of dogs on rather low diet. They seem to have had an oval tent. We sleep one night at the Pole and have had a double hoosh with some last bits of chocolate, and Ber's[17] cigarettes have been much appreciated by Scott and Oates and Evans. A tiring day — now turning in to a somewhat starchy frozen bag. Tomorrow we start for home and shall do our utmost to get back in time to send the news to the ship.

Thurs 18 Jan   Sights were taken in the night and at about 5 a.m. we turned out and marched from this night camp about 3¾ miles back in a S. Ely direction to a spot which we judged from last night's

1910—JUNE

22. ○ Full Moon 8.12 p.m.
24. Midsummer Day

NAME AND ADDRESS	20	21	22	23	24	25	26	£	s.	d.
	M	T	W	Th	F	S	S			

[handwritten diary entry, Wednesday 17 Jan]

A page from Wilson's sledging diary: Wednesday 17 Jan

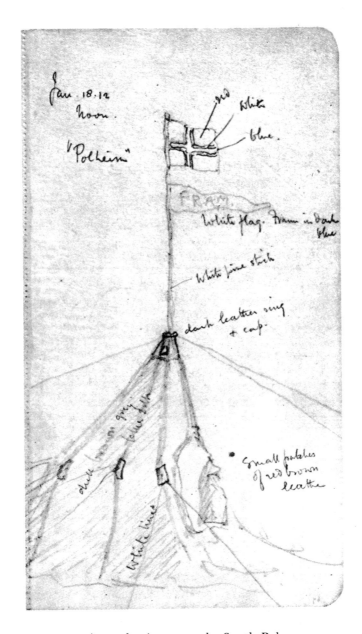

Amundsen's tent at the South Pole

sights to be the Pole. Here we lunched camp — built a cairn — took photos, flew the Queen Mother's Union Jack and all our own flags. We call this the Pole, though as a matter of fact we went half a mile further on in a S. Ely direction after taking further sights to the actual final spot and here we left the Union Jack flying. During the forenoon we passed the Norwegians' last southerly camp. They called it Polheim and left here a small tent with Norwegian and *Fram* flags flying and a considerable amount of gear in the tent, half reindeer sleeping bags, sleeping socks, reinskin trowsers 2 pair, a sextant and artificial horizon, a hypsometer with all the thermometers broken etc. I took away the spirit lamp of it which I have wanted for sterilizing and making disinfectant lotion of snow. There were also letters there. One from Amundsen to King Haakon with a request that Scott should send it to him. There was also a list of the 5 men who made up their party, but no news as to what they had done. I made some sketches here but it was blowing very cold −22. Birdie took some photos. We found no sledge there though they said there was one. It may have been buried in drift. The tent was a funny little thing for 2 men, pegged out with white line and tent pegs of yellow wood. I took some strips of blue grey silk off the tent seams. It was perished. The Norskies had got to the Pole on December 16 and were here from 15th to 17th. At our lunch South Pole camp we saw a sledge runner with a black flag about half a mile away bearing from it. Scott sent me on ski to fetch it and I found a note tied to it showing that this was the Norskies actual final Pole position. I was given the flag and the note with Amundsen's signature and I got a piece of the sledge runner as well. The small chart[18] of our wanderings shows best how all these things lie. After lunch we made 6.2 miles from the Pole camp to the north again — and here we are camped for the night.

Fri 19 Jan   We followed this course till we struck the small cairn where we lost Amundsen's tracks — here we picked up our own and his and followed them N.E. till we reached his next small cairn and then the black flag camp our 68th outward camp. We went on due N. from here — taking his stick and flag — and lunched at 1 having done 8.1 miles. We have the floorcloth spread as a sail on the tent poles and inner tent as a mast. We had a splendid wind right behind us most of the afternoon and went well until about 6 p.m. when the sun came out and we had an awful grind until 7.30 when we camped. The sun comes out on sandy drifts all on the move in the wind and temp. −20° and gives us an absolutely awful surface with no glide at all for ski or sledge and just like fine sand. The weather all day has been more or less overcast with white broken alt. str.[19] and for 3

Amundsen's South Pole mark

degrees above the horizon there is a grey belt looking like a blizzard of drift, but this in reality is caused by a constant fall of minute snow crystals — *very* minute — sometimes instead of crystal plates the fall is of minute agglomerate spicules like tiny sea urchins. The plates glitter in the sun as though of some size, but [you] can only just see them as pin points on your. burberry. So the spicule collections are also only just visible. Our hands are never warm enough in camp to do any neat work now the weather is always uncomfortably cold and windy about −23, but after lunch today I got a bit of drawing done. Sastrugi today definitely S. Wly and S. Ely. Our old tracks are already covered in places by 1 foot old looking sastrugi — all fresh, many deeply cut, but not hard. We passed 67½ camp this afternoon and are now 9 miles from the last outward depot which is 66½ camp.

Sat 20 Jan   Blowing 4−5 from the S.S.W. and from time to time overcast but too much sun and moving sandy drifts at this temp. for good going — they refused to allow the ski to glide at all. One had to shove them over. The going was irregular and uncomfortable but we made 9.3 miles by lunch in 5 hours, 10 minutes, when we reached 66½ camp where we had left a depot 'The last depot'. We passed camp 67 in the forenoon march. This afternoon we ought to pass camp 66 at about 7 miles from this depot. Snow was falling all forenoon again in minute little collections of spicules, but made a very grey mist all round and gave a complete double 22° halo. The lower edge of the inner ring just touching the horizon. One has to wear snow goggles always up here especially now that we are walking into the sun. We picked up a bamboo here used as our depot flag and discarded Amundsen's 'Pole' — from which I took a number of hickory splinters. We go on from this depot with just 7 days' provision for the next 50 miles when we pick up our 1½ degree depot. In the afternoon we had a very heavy pull until 7.30 passing camp 66 shortly before halting. The weather very thick with a great quantity of drifting surface sandy snow which made drifts that clung to our ski like glue. Camp 66 = night of 14th Jan. The wind all day has been S.S.W. and a fine double halo all day with horns as before. Bags and gear getting a bit wet and frozen up. No chance to dry things for some time now and temperature low, −28 last night, −18 this evening, always drift and snow crystals and wind and not much sun. Evans has got 4 or 5 of his finger tips badly blistered by the cold. Titus also his nose and cheeks — so Evans and Bowers.

Sun 21 Jan   We had a blizzard −18 to −11 and so thick that we had to lie up for the forenoon. It blew about 6, and it would have been impossible to follow the tracks, but after lunch the sun came out and it suddenly cleared and the wind died down blowing from S. by W. We got away at a quarter to 4 and marched till 7.45 and did 5.5 miles passing one cairn no. 65½. We are now about 89° 14′ S. and have 6 days food in hand to next depot. The cairn we passed was built of large single blocks of snow and had leaned down to 45° without any falling off.

Mon 22 Jan   Sun out and temp. as usual with clearing sky is dropping. Crystals in the air all the same but fewer and one could see cairn 65 about 1½ miles away today. We marched from 8 a.m. to 1 with a really heavy surface and very little breeze indeed. We made 8.3 miles geog. The S.S.E. sastrugi which we crossed coming over this march on our way S. are now all replaced by deep cut S. by Wly sastrugi in abundance. After lunch we made 6.2 miles and passed camp 64½ and went on from 3 p.m. to 7 p.m. — a very heavy drag and no wind but sun out. Temp. −21. We are apparently going over

151

very gradual undulations, but the aneroid does not show much change. We are now 10,000 ft above sea level and the Pole 9,500. We have to rise to 10,500 before we begin going down the Beardmore. We are now 30 miles from our next depot and ought when we get there to have 10 days' food for the next 95 miles to the next depot.

Tues 23 Jan   We started for 2 miles in hot sun with no wind — then very heavy pulling — then wind came on much harder and drift, and we got off the tracks constantly. I wrote this at lunch and in the evening had a bad attack of snow blindness.

Wed 24 Jan   Blizzard in afternoon. We only got in a forenoon march. Couldn't see enough of the tracks to follow at all. My eyes didn't begin to bother me till tomorrow though it was the strain of tracking and the very cold drift which we had today that gave me this attack of snow glare.

Thurs 25 Jan   Marched on foot in the afternoon as my eyes were too bad to go on on ski. We had a lot of drift and wind and very cold. Had Zn SO$_4$ and cocaine in my eyes at night and didn't get to sleep at all for the pain — dozed about an hour in the morning only.

Fri 26 Jan   Marched on foot again all day as I couldn't see my way on ski at all. Birdie used my ski. Eyes still very painful and watering. Tired out by the evening — had a splendid night's sleep — and though very painful across forehead to light they are much better.

Sat 27 Jan   Very bad surface of deep cut sastrugi all day until late in the afternoon when we began to get out of them. We passed cairns 59 and 58½ and 58. One of these had curved into a semi-circle without dropping a block. Eyes better, was on ski all day again.

Sun 28 Jan   We had a fine day and a good march on very decent surface — a few isolated deep sastrugi, but otherwise nearly all surface wind marks like butterfly scales. Made sail as usual and had a brisk breeze most of the afternoon. Titus picked up his lost pipe at our camp tonight which was the lunch camp of Jan 4th. It was just showing. We are camped at this cairn tonight. My eyes are well again and we are all on ski except Birdie. We are about 10,130 ft above s.l. We are now about 42 miles from our 3 degree depot, the next one, and when we get there we ought to have 3 days' food in hand to increase on. We are all pretty hungry — could eat twice what we have, especially at lunch and breakfast. Evans has a number of badly blistered finger ends which he got at the Pole. Titus' big toe is turning blue black. Lat. tonight about 87°39'.

Mon 29 Jan We got in a very long march for 9 hours going over part good and part bad surface. There has been a lot of very glassy porcelain shell surface with raised footprints and sledge tracks on it and enormous snowdrifts and banks of hard crusted very deep cut sastrugi, awful for skiing over, but Scott and I were on ski the whole day — the other three on foot. I got a nasty bruise on the Tib. ant. which gave me great pain all the afternoon. The sky radiating windy cirrus S.E. and N.W. and a stiff S.E. wind with low drift all day. Sky cleared at night. Temp. −25 with it made it very cold. We are now only 22 miles from our depot and 400 miles about to go before meeting the dogs with ship's news. Tonight about 87°20'. We passed the cairn of the last camp we had with the 2nd supporting party in the forenoon and from there onward had 3 sledge tracks to follow.

Tues 30 Jan My left leg exceedingly painful all day so I gave Birdie my ski and hobbled alongside the sledge on foot. The whole of the Tibialis anticus is swollen and tight and full of tenosynovitis and the skin red and œdematous over the shin. But we made a very fine march with the help of a brisk breeze and a good going surface. Shell porcelain and very high deep cut sastrugi in patches under and over the old tracks.

Wed 31 Jan Again walking by the sledge with swollen leg but not nearly so painful. We had 5.8 miles to go to reach our three degree depot, picked this up with a week's provision and a line from Evans and then for lunch an extra biscuit each, keeping 4 for lunch and 1/10 whack of butter extra as well. Afternoon we passed cairn where Birdie's ski had been left. These we picked up and came on till 7.30 p.m. when the wind which had been very light all day dropped and in temp. −20 it felt delightfully warm and sunny and clear. We have 1/10 extra pemmican in the hoosh now also. My leg pretty swollen again tonight. Evans' finger nails all coming off, very raw and sore. Surface is still biscuit porcelain with large sastrugi above and below the old tracks.

Thurs 1 Feb Cairns 2 in 6.5 and 5.5 miles were to be passed today, but we passed them and camped at the last having covered 15.7 miles. Temp. −20 in afternoon, a stiff breeze and porcelain eggshell china surface alternating with huge deep cut sastrugi. My

153

leg much more comfortable — gave me no pain and I was able to pull all day holding on to the sledge. Still some œdema. We came down a hundred feet or so today on a fairly steep gradient.

Fri 2 Feb   We came down two slopes today. Scott had a nasty fall on the point of his shoulder. We have not however seen a crevasse as yet on the way back but ought to tomorrow. We are still following our old tracks. I got through the day without any slips and no further damage to leg which is mending well. The surface was very largely eggshell porcelain, white and glistening like a woodpecker's egg and as thin — almost always gives way under foot. [In] some places there are 6 or 7 thin crusts in 2 or 3 inches depth of this snow. Wind was off and on today — very cold forenoon for me in the sail's shadow, but very warm afternoon march. Finished at 8 p.m. Very tired — sleep at once.

Sat 3 Feb   Sunny and breezy again. Came down a series of slopes and finished the day by going up one. Enormous deep cut sastrugi and drifts and shiny eggshell surface. Wind all S.S. Ely. Today at about 11 p.m. we got our first sight again of mountain peaks on our eastern horizon almost abeam as we went N. about E. by N. or E.N.E. We had 9 hours of marching and finished at 8 p.m. Sunny and very little S. Ely breeze. We crossed the outmost line of crevassed ridge top today — the first on our return. Lost the old tracks and are now going due north — with still a week's food for about 65 miles to the Mt. Darwin depot.

Sun 4 Feb   We had 9 hours marching with a steady useful S.S. Ely breeze all day. All the others went on ski in the afternoon but we all foot slogged all forenoon. Temp. —23. Clear cloudless blue sky — surface drift. During forenoon we came down gradual descent including 2 or 3 irregular terrace slopes on crest of one of which were a good many crevasses. Southernmost were just big enough for Scott and Evans to fall in to their waists — and very deceptively covered up. They ran east and west. Those nearer the crest were the ordinary broad street-like crevasses well lidded. In the afternoon we again came to a crest before descending with street crevasses and one we crossed had a huge hole where the lid had fallen in, big enough for a horse and cart to go down. We have a great number of mountain tops on our right and south of our beam as we go due north now. We are now camped just below a great crevassed mound on a mountain top evidently. The surface has been marble smooth on the crests with temperature cracks and on the south side rather softer butterfly

154

scale. On the northern sides which are steeper the sastrugi are immense and deeply cut and often hard. Since the last depot we have been having extra food, making a week last 6 days, I believe, and we are grateful for it, 4 biscuits for lunch and fat hooshes always now. Evans is feeling the cold a lot always getting frost bitten. Titus' toes are blackening and his nose and cheeks are dead yellow. Dressing Evans' fingers every other day with boric vaseline — they are quite sweet still.

Mon 5 Feb   We had a difficult day getting in among a frightful chaos of broad chasm-like crevasses. We kept too far east and had to wind in and out amongst them and cross a multitude of bridges. We then bore west a bit and got on better all the afternoon and got round a good deal of the upper disturbance of the falls here. We ought to have gone west of the big hump (see yesterday) and we should have missed all today's trouble. The weather was perfect — cloudless blue sky and sun [illegible] breeze most of the day and none at camping. We camped for the night on hard snow among crevasses Evans' fingers suppurating, nose very bad and rotten looking. Land well in sight today. Mt. Darwin, Buckley, Dominion Range, Flat Tops, Kirkpatrick etc. The falls are all formed of large open crevasses.

Tues 6 Feb   We again had a forenoon of trying to cut corners. Got in amongst great chasms running E. and W. and had to come out again. We then again kept west and down hill over tremendous sastrugi with a slight breeze very cold — and afternoon continued bearing more and more towards Mt. Darwin. We got round one of the main lines of icefall and looked back up to it. We are now camped about 10 miles from the Upper Glacier Depot, but have seen none of our old tracks or cairns. We are on a terrace with another good drop ahead for tomorrow. Dominion Range has been looking very fine today — close enough to see some colour in the rock. The breeze rose again in the night Sly. Blew all day and fell calm — 15 in the evening. Very cold march — many crevasses and walking by the sledge on foot found a good many — the others on ski. Weather last night became completely overcast with alto stratus, but all cleared quickly in the morning and we had a cloudless day of sunshine. Periodicity of Sly and S. Ely breezes seems possible — it often falls in the evening to nil and then feels real warm.

Wed 7 Feb   Clear day again and we made a tedious march in the forenoon along a flat or two and down a long slope and then in the afternoon we had a very fresh breeze and very fast run down last slopes covered with big sastrugi. It was a strenuous job steering and checking behind by the sledge. We reached the Upper Glacier Depot by 7.30 p.m. and found everything right which was satisfactory, after a breakfast which was given up to a discussion as to the absence of one day's biscuit. The colour of the Dominion Range rock is in the main all brown madder or dark reddish chocolate — but there are numerous narrow bands of yellow rock scattered amongst it. I think it is composed of dolerite and sandstone as on the W. side.

Thurs 8 Feb   A very busy day. We had a very cold forenoon march blowing like blazes from the S. Birdie detached and went on ski to Mt. Darwin and collected some dolerite, the only rock he could see on the nunatak[20] which was nearest. We got into a sort of crusted surface where the snow broke through nearly to our knees and the sledge runner also. I thought at first we were all on a thinly bridged crevasse. We then came on east a bit and gradually got worse and worse going over an icefall having great trouble to prevent sledge taking charge, but eventually got down and then made N.W. or N. into the land and camped right by the moraine under the great sandstone cliffs of Mt. Buckley, out of the wind and quite warm again — was a wonderful change. After lunch we all geologised until supper, and I was very late turning in, examining the moraine after supper. Socks all strewn over the rocks dried splendidly. Magnificent Beacon Sandstone cliffs. Masses of limestone in the moraine — and dolerite crags in various places. Coal seams at all heights in the sandstone cliffs and lumps of weathered coal with fossils, vegetable. Had a regular field day and got some splendid things in the short time[21].

Fri 9 Feb   We made our way along down the moraine and at the end of Mt. Buckley unhitched and had half an hour over the rocks and again got some good things written up in sketch book. We then left the moraine and made a very good march on rough blue ice all day with very small and scarce scraps of névé on one of which we camped for the night with a rather overcast foggy sky which cleared to bright sun in the night. We are all thoroughly enjoying temps. of + 10 or there-about now with no wind instead of the summit winds which are incessant with temp. —20.

Sat 10 Feb   We made a very good forenoon march from 10 to 2.45 towards the Cloudmaker. Weather overcast gradually obscured everything in snowfull fog, starting with crystals of large size — plate stars beautifully shaped and size of this —

changed to smaller of the same gradually by 7.30 p.m., when we camped, and after hoosh to minute agglomeration of spicules. We had to camp after 2½ hours afternoon march as it got too thick to see anything and we were going downhill on blue ice after crossing a few inches of névé. The snow in this névé is pitted in small cups on the east and S.E. but sastrugized with no pitting on the N. and N.W. This seems funny as the pitting is evidently due to the sun. An immense amount of rock dust is blown over the glacier and catches in sastrugi and in the cracks and snow-filled crevasses — gets stuck and makes the snow more sticky and goes on increasing so. Also water holes from these collections into the crevasses — but no water seen. These evidently form the silt bands — but there are bands of older dirty ice, possibly moraine sunk, which run down the glacier often at right angles to all the tension cracks and crevasses. These are old and planed off level with surface and often faulted. We are tonight about 20 − 25 miles from the Mid-glacier Depot and are making for the Cloudmaker. We have Wild Mt. and a glacier and Marshall Mt. on our left beam and have dropped Plunket P. today on our right quarter.

Mon 12 Feb   We had a good night just outside the icefalls and disturbances and a small breakfast of tea, thin hoosh and biscuit, and began the forenoon by a decent bit of travelling on rubbly blue ice on crampons — then plunged into an icefall and wandered about in it absolutely lost for hours and hours.

Tues 13 Feb   We had one biscuit and some tea after a night's sleep on very hard and irregular blue ice amongst the icefall crevasses — no snow on the tent, only ski etc. Got away at 10 a.m. and by 2 p.m. found the depot having had a good march over very hard rough blue ice, only half an hour in the disturbance of yesterday. The weather was very thick, snowing and overcast. Could only just see the points of bearing for depot. However we got there, tired and hungry, and camped, and had hoosh and tea and 3 biscuits each. Then away again with our 3½ days of food from this red flag depot and off down by the Cloudmaker moraine. We travelled about 4 hours on hard blue ice and I was allowed to geologise the last hour down the two outer lines of boulders. The outer are all dolerite and quartz rocks — the inner all dolerite and sandstone. The Cloudmaker has a huge bank of moraine terrace down half (lower) of its length, 700 ft above level of glacier. South of Cloudmaker is a glacier and S. of it another great cape of moraine.
We camped on the inner line of boulders — weather clearing all the afternoon.

Wed 14 Feb   We made a good day's march along the ridge of a very long pressure ridge. I was on foot and the rest all on ski. We passed one boulder only, one of agglomerate sandstone made up of many coarse pebbles definitely waterworn of all sizes.

Thurs 15 Feb   I got on ski again, first time since damaging my leg, and was on them all day for 9 hours. It was a bit painful and swelled by the evening and every night I put on [a] snow poultice. We are not yet abreast of Mt. Kyffin and much discussion how far we are from the Lower Glacier Depot — probably 18 to 20 miles — and we have to reduce food again, only one biscuit tonight with a thin hoosh of pemmican. Tomorrow we have to make one day's food which remains last over the two. The weather became heavily overcast during the afternoon and then began to snow and though we got in our 4 hours' march it was with difficulty — and we only made a bit over 5 miles. However, we are nearer the depot tonight.

Fri 16 Feb   Got a good start in fair weather after one biscuit and a thin breakfast and made 7½ miles in the forenoon. Again the weather became overcast and we lunched almost at our old bearing on Kyffin of lunch Dec. 15th. All the afternoon the weather became thicker and thicker and after 3¼ hours Evans collapsed — sick and giddy and unable to walk even by the sledge on ski, so we camped. Can see no land at all anywhere but we must be getting pretty near the Pillar Rock. Evans' collapse has much to do with the fact that he has never been sick in his life and is now helpless with his hands frost-bitten. We had thin meals for lunch and supper.

Sat 17 Feb   The weather cleared and we got away for a clear run to the depot, and had gone a good part of the way when Evans found his ski shoes coming off. He was allowed to readjust and continue to pull, but it happened again and then again, so he was told to unhitch, get them right and follow on and catch us up. He lagged far behind till lunch and when we camped we had lunch and then went back for him as he had not come up. He had fallen and had his hands frost-bitten and we then returned for the sledge and brought it and skid him in on it, as he was rapidly losing the use of his legs. He was comatose when we got him into the tent and he died without recovering consciousness that night about 10 p.m. We had a short rest for an hour or two in our bags that night, then had a meal and came on through the pressure ridges about 4 miles further down and reached our Lower Glacier Depot. Here we camped at last, had a good meal and slept a good night's rest which we badly need. Our depot was all right.

The Pillar Rock near where Petty Officer Evans died on 17 February 1912. Sketched on 11 December 1911

Sun 18 Feb   We had only 5 hours' sleep. We had butter and biscuit and tea when we woke at 2 p.m. then came over the gap entrance to the pony slaughter camp, visiting a rock moraine of Mt. Hope on the way.

Mon 19 Feb   Late in getting away after making up new 10 foot sledge and digging out pony meat. We marched 5½ miles was on very heavy surface indeed.

Tues 20 Feb -   We got to the Blizzard Camp[22] by lunch time.

Wed 21 Feb   We had a solid day's march and picked up one or two things — the horse walls of Dec. 3. and an old tent ring and some ski tracks of returning parties. We had a very heavy surface of deep soft snow and we made 8½ miles in the day on ski.

Thurs 22 Feb   Soon after starting it came on to blow and surface drift from the S.S.E. We were on ski and got the sail up and made 5½ miles in the forenoon. We failed to see anything of either the cairn of Dec. 2 or the horsewalls of Dec. 2, where Victor was killed. We had a great pony hoosh in the evening. The wind lasted all day but fell in the evening.

Fri 23 Feb   Bad day off the tracks. We had thick weather and got no cairns. Lost much time in discussing navigation. We were too far out and had to come well in. The drift and sunshine gave me very bad eyes.

Sat 24 Feb   Bad attack of snowglare — could hardly keep a chink of eye open in goggles to see to the course. Fat pony hoosh.

Sun 25 Feb   My eyes much better. Started my week of cooking. No time for anything at any meals. Very good day's going on ski. Took on job of pace maker and got sweated through — very cold night.

Mon 26 Feb   Good day's going on ski with little breeze from S.S.E. Fat pony hoosh. Temp down to −37 in the night.

Tues 27 Feb   Overcast all forenoon and cleared to splendid clear afternoon. Good march on ski. Some fair breeze. Turned in at −37,

EPILOGUE

Here the diary ends. From this point onward conditions on the march for the Pole Party steadily worsened. When Mid Barrier Depot was reached on 1 March a critical shortage of fuel was discovered. On or about 16 March Oates, suffering terribly from frostbitten feet, walked out of the tent to his death.

The three survivors struggled on until 21 March when, after a nine day blizzard, they died in their tent, eleven miles from One Ton Depot and safety. Their bodies were discovered eight months later by a search party led by Surgeon Atkinson.

Following page: *The Barrier Silence:* published in *The South Polar Times*, October 1911. According to George Seaver, Wilson's biographer, it was Wilson's 'first and only poem'

THE Silence was deep with a breath like sleep
        As our sledge runners slid on the snow,
And the fate-full fall of our fur-clad feet
        Struck mute like a silent blow
On a questioning "hush", as the settling crust
        Shrank shivering over the floe;
And the sledge in its track sent a whisper back
        Which was lost in a white fog-bow.

AND this was the thought that the Silence wrought
        As it scorched and froze us through,
Though secrets hidden are all forbidden
        Till God means man to know,
We might be the men God meant should know
        The heart of the Barrier snow,
        In the heat of the sun, and the glow
        And the glare from the glistening floe,
As it scorched and froze us through and through
        With the bite of the drifting snow.

BIOGRAPHICAL NOTES

Members of Scott's *Discovery* and *Terra Nova* expeditions mentioned in Wilson's diary extracts.

Albert B. ARMITAGE (1864-1943)
Nicknamed 'The Pilot'. Second-in-command of the Jackson-Harmsworth expedition, 1894-1897, to Franz-Josef Land. During the *Discovery* expedition he was second-in-command and navigator. His own narrative of the expedition was published in 1905 under the title *Two years in the Antarctic*.

Edward Leicester ATKINSON (1882-1929)
Surgeon R.N. known as 'Atch'. He joined Scott's *Terra Nova* expedition as parasitologist and bacteriologist. He was in command during the last year at Cape Evans and succeeded in the task of maintaining morale during the difficult time that followed the finding of the bodies of Scott and his companions.

Michael BARNE (1877-1961)
Second lieutenant R.N. on *Discovery*. He acted as assistant magnetic observer.

Louis C. BERNACCHI (1876-1940)
Served as magnetic and meteorological observer during the *Southern Cross* expedition, 1898-1900. Joined *Discovery* in New Zealand as physicist.

Henry Robertson BOWERS (1883-1912)
Known as 'Birdie' on account of his beak-like nose. Originally joined the *Terra Nova* expedition as a ship's officer in charge of stores, but Scott was so impressed with his organizational ability that he decided to keep him with the shore party. An expert navigator and the toughest member of the expedition. He died with Scott and Wilson on the Barrier towards the end of March 1912.

Apsley George Benet CHERRY-GARRARD (1886-1959)
Known as 'Cherry'. He was introduced to Scott by Wilson and joined

the expedition as assistant zoologist though he had no scientific qualifications. He proved to be highly versatile, edited the expedition's magazine *The South Polar Times* and took part in all the major sledge journeys. In March 1912, with the Russian dog driver, Anton Omelchenko, he tried to relieve Scott's returning party from the South Pole. His failure to do so, though no fault of his own, preyed on his mind for the rest of his life. In 1922 his account of the expedition, *The worst journey in the world*, was published, probably the best account of a polar expedition ever written.

Thomas CREAN (died 1938)
Petty Officer R.N. A native of County Kerry, Ireland. Served on both the *Discovery* and *Terra Nova* expeditions. Later served as second officer on Shackleton's *Endurance* expedition of 1914-1917 and was a member of the *James Caird* boat party from Elephant Island to South Georgia.

Jacob CROSS
Petty Officer R.N. on *Discovery*. Helped Wilson during the expedition with bird skinning.

Fred E. DAILEY
Carpenter and Warrant Officer R.N.

Bernard DAY
Before joining Scott's *Terra Nova* expedition as motor engineer he had been in charge of the motor car used on Shackleton's *Nimrod* expedition of 1907-1909.

Demitri *see* Demetri GEROV

Gerald DOORLY
Junior officer on the *Discovery's* relief ship *Morning*.

Edgar EVANS (1876-1912)
Petty Officer, R.N. Known as 'Taff'. Born at Rhossili, Wales. Joined the Royal Navy in 1891 and in 1901 volunteered for service with Scott's *Discovery* expedition. On his return to England he became a gunnery instructor. Though Evans occasionally fell from grace Scott thought the world of him: 'A giant worker with a really remarkable headpiece.'

Edward Ratcliffe Garth Russell EVANS, *afterwards* LORD MOUNTEVANS (1881-1957)
Sometimes known as 'Teddie'. Lieutenant R.N. on Scott's *Terra*

Nova expedition which he joined as navigator and second-in-command. He had previously served as a sub-lieutenant on the *Morning*, which, with the *Terra Nova*, relieved the *Discovery* in 1902. In the Antarctic he was leader of the last supporting party to leave Scott on the journey to the South Pole. During the return journey he suffered acutely from scurvy and would have died had it not been for the efforts of his two companions, William Lashly and Thomas Crean (*q.v.*) to sledge him back to base. He was invalided home in 1912 but returned in the *Terra Nova* in January 1913 to take charge of the expedition. His account of Scott's last expedition was first published in 1921 as *South with Scott*.

Thomas A. FEATHER
Warrant Officer R.N. Served on the *Discovery* as boatswain.

Demetri GEROV (1888?-1932)
Dog driver on Scott's *Terra Nova* expedition. A native of Sakhalin, eastern Siberia, he was chosen by Cecil Meares to help pick the dogs in Nikolayevsk and accompany them to Antarctica.

Tryggve GRAN (1889-1980)
Sub-lieutenant in the Norwegian Naval Reserve. He joined the *Terra Nova* as a ski expert having been introduced to Scott by the great Norwegian explorer Fridtjof Nansen. Gran was one of the first of the relief party to sight Scott's tent in November 1912.

T.V. HODGSON (1864-1926)
Marine biologist on the *Discovery* expedition and afterwards Curator of Plymouth Museum.

Frederick J.HOOPER (1891-1955)
Petty Officer R.N. Joined the *Terra Nova* expedition as a steward, but was transferred to the Shore Party where he proved a valuable member of the expedition. He was a member of the search party which found the bodies of Scott and his companions on 12 November 1912.

Patrick KEOHANE
Petty Officer R.N. An Irishman from County Cork. With Wright, Atkinson and Cherry-Garrard he was a member of Scott's supporting party which turned back at the top of the Beardmore Glacier on 21 December 1911.

Reginald KOETTLITZ (1861-1916)
Surgeon and botanist on the *Discovery* expedition. Previously surgeon on the Jackson-Harmsworth expedition to Franz-Josef Land, 1894-1897.

William LASHLY (died 1940)
Chief Stoker R.N. Served with Scott on both the *Discovery* and *Terra Nova* expeditions and proved himself to be a first-rate sledger and all-rounder. Accompanied the Pole Party as far as the polar plateau. With Petty Officer Crean he succeeded in sledging Lieutenant Evans safely back to base after the latter was stricken with scurvy.

Cecil H.MEARES (died 1937)
Was responsible for the purchase and management of the dogs on the *Terra Nova* expedition.

J.D.MORRISON
Chief Engineer of the *Discovery's* relief ship *Morning*.

George MULOCK (1882-1963)
Third Lieutenant and Surveyor R.N. Replaced Shackleton when he was invalided home on the *Morning* in 1903. Was responsible for the compilation of the survey on the *Discovery* expedition.

Edward W.NELSON
Invertebrate biologist on the *Terra Nova* expedition.

Laurence Edward Grace OATES (1880-1912)
Captain, 6th Inniskilling Dragoons. Known as 'Titus', or sometimes as 'The Soldier'. Volunteered to take charge of the ponies on Scott's *Terra Nova* expedition. Was selected for the Pole Party by Scott, probably to represent the Army. The famous painting by J.C. Dollman of Oates going to meet his death in the blizzard — 'A very gallant gentleman' — now hangs in the Cavalry Club in London.

Charles W. Rawson ROYDS (1876-1931)
First Lieutenant on *Discovery*. He was in charge of the expedition's meteorological work as well as supervising the work of the men and the internal economy of the ship.

Robert Falcon SCOTT (1868-1912)
Called 'The Owner'. In 1899 while serving on H.M.S. *Majestic* he was informed by Sir Clements Markham, President of the Royal Geographical Society, of his plans for an Antarctic expedition to be organized in co-operation with the Royal Society. Scott applied for its command and was appointed. In 1901 he sailed south on the *Discovery*, returning to England in 1904 having carried out

successfully the first extensive land exploration of the Antarctic continent. He was promoted Captain the same year. In 1908 he married Kathleen Bruce by whom he had a son Peter Markham (the present Sir Peter Scott). He returned to Antarctica in 1910 at the head of his own expedition determined to continue the scientific work started on the *Discovery* expedition, and also to achieve the South Pole. With Wilson, Bowers, Oates and Petty Officer Evans he reached the Pole on 18 January 1912 only to find that a Norwegian expedition led by Roald Amundsen had been there a month previously. All five members of the Pole Party perished on the return journey.

Ernest Henry SHACKLETON (1874-1922)
Sub-Lieutenant R.N.R. on the *Discovery*. Nicknamed 'Shackle'. He entered the Merchant Service in 1890 and served with Union Castle Line before joining the expedition. He was in charge of seawater analysis and also edited the expedition's magazine *The South Polar Times*. Invalided home in 1903. He subsequently led three expeditions of his own to the Antarctic, in 1907-1909 (*Nimrod*) 1914-1917 (*Endurance*) and 1921-1922 (*Quest*). On this last expedition he died and was buried on South Georgia in 1922. He was knighted in 1909.

George Clarke SIMPSON (1878-1965)
Known as 'Sunny Jim'. Meteorologist on the *Terra Nova* expedition. Later he became head of the Meteorological Office, London, and was knighted in 1935.

Reginald SKELTON (1872-1952)
Chief Engineer R.N. on the *Discovery* whose construction at Dundee he supervised. Acted as photographer on the expedition. Later he was closely involved with the designing of the motor sledges for Scott's *Terra Nova* expedition.

Thomas Griffith TAYLOR (1880-1964)
Known as 'Griff'. Educated in Australia he joined the *Terra Nova* expedition as physiographer. His own account of the expedition *With Scott: the silver lining* was published in 1915.

Charles Seymour WRIGHT (1887-1975)
Nicknamed 'Silas'. A native of Toronto, he was the only Canadian on Scott's *Terra Nova* expedition which he joined as physicist. He was a member of the search party in November 1912 and was the first to sight Scott's tent drifted up with snow. In later life he became Director of Scientific Research at the Admiralty and was knighted in 1946.

FURTHER READING

ARMITAGE, Albert B., *Two years in the Antarctic*, Edward Arnold, London, 1905.

CHERRY-GARRARD, Apsley, *The worst journey in the world*, Constable, London, 1922 (and subsequent editions by Chatto & Windus).

EVANS, E.R.G.R., *South with Scott*, Collins, London, 1921.

HUXLEY, Elspeth, *Scott of the Antarctic*, Weidenfeld and Nicolson, London, 1977.

KING, H.G.R., *The Antarctic*, Blandford Press, London, 1969.

KIRWAN, L.P., *A history of polar exploration*, Penguin Books, London, 1962.

PONTING, H.G., *The great white south*, Duckworth, London, 1921.

ROBERTS, B.B., *Edward Wilson's birds of the Antarctic*, Blandford Press, London, 1967.

SCOTT, R.F., *The voyage of the Discovery*, Vols 1 and 2, Macmillan, London, 1905.

SCOTT, R.F., *Scott's last expedition, being the journals of Captain R.F. Scott, R.N., C.V.O., arranged by Leonard Huxley*, Vols 1 and 2, Smith, Elder, London, 1913.

SEAVER, George, *Edward Wilson of the Antarctic*, John Murray, London, 1934.

SEAVER, George, *Edward Wilson: nature-lover*, John Murray, London, 1937.

SEAVER, George, *The faith of Edward Wilson*, John Murray, London, 1948.

SOUTH POLAR TIMES, THE, Volumes 1 and 2, Smith, Elder, London, 1907, Volume 3, Smith, Elder, London, 1914.

TAYLOR, T. Griffith, *With Scott: the silver lining*, Smith, Elder, London, 1916.

WILSON, Edward Adrian, *Diary of the 'Discovery' expedition to the Antarctic regions 1901-1904*, Edited from the original mss. in the Scott Polar Research Institute, Cambridge, by Ann Savours, Blandford Press, London, 1966.

WILSON, Edward Adrian, *Diary of the 'Terra Nova' expedition to the Antarctic 1910-1912*, An account of Scott's last expedition edited from the original mss. in the Scott Polar Research Institute and the British Museum by H.G.R. King, Blandford Press, London, 1972.

NOTES

THE SOUTHERN JOURNEY, SUMMER 1902-1903

1 Davos, Switzerland, where Wilson recuperated from tuberculosis in 1898.
2 Lieutenant Albert B. Armitage known as 'The Pilot'. *See* biographical notes.
3 Wave-like dunes, up to six feet in height, usually lying in the direction of the prevailing wind.
4 Reindeer-skin boots lined for warmth with a dried grass.
5 Warrant Officer Thomas A. Feather, R.N.
6 Now known as McMurdo Sound.
7 Director of the Cape Town Observatory.
8 Afterwards named Shackleton Inlet.
9 Afterwards named Cape Wilson.
10 Robert Edwin Peary (1856-1920), American explorer who sledged extensively over northeast Greenland in an attempt to reach the North Pole. He finally achieved his ambition on 6 April 1909.
11 Shackleton is believed to have suffered from a weakness of the heart resulting from rheumatic fever in childhood. In this instance the condition may have been brought on by hauling the heavy sledge.

THE WORST JOURNEY IN THE WORLD

1 Presumably refers to Otto Sverdrup's expedition to the Canadian Arctic on the *Fram*, 1898-1902.
2 The Owner — a slang term for the captain of a ship — Scott.
3 No figure given. Nowadays ration values are calculated in calories, or more properly kilocalories, and not in foot pounds of energy.
4 The anniversary of Wilson's wedding.
5 Wilson typically plays down what must have been an agonizing experience. In his official report he deleted all mention of the incident.

SLEDGE JOURNEY TO THE SOUTH POLE

1 Holy Communion. It was Wilson's custom to celebrate a private communion service on the march.
2 Blossom — one of the ponies who died on the depot-laying party in February 1911.
3 Cirrostratus — cloud formation.
4 Blucher — another pony who collapsed on the depot-laying party.
5 Geographical miles (1 mile geographical = 1.15 statute miles).
6 On 21 November 1908, just south of the 81st parallel.
7 Not reproduced here.
8 The Gap, or 'Gateway' as Shackleton's 1907-1909 expedition named it, is the passage between Mount Hope and the mainland leading to the Beardmore Glacier.
9 A geological term referring to a distinctive group of fossil-bearing sedimentary formations discovered during Scott's *Discovery* expedition.
10 Shackleton's position at midday on 8 December 1908.
11 This glacier is provisionally identified and described with a memory sketch in one of Wilson's sketchbooks.
12 Left blank in the manuscript. Mount Wild seems the most probable landmark.
13 Presumably sheets of thin brittle ice.
14 Two hoops = first rate.
15 Stratocumulus cloud.
16 With this entry Wilson occasionally writes the camp number in the margin of the diary. Camp 1 was Hut Point.
17 Wilson's brother Bernard.
18 Chart not found.
19 Altostratus cloud.
20 Nunatak — a rocky crag or small mountain projecting through the ice.
21 Wilson's notes on this visit to Mount Buckley occupy six pages in one of his two sketch books. The rocks he collected are now in the British Museum (Natural History).
22 Camp of 5—8 December 1911.

INDEX

Numbers in *italic* refer to illustrations. Where appropriate sub-entries have been given in chronological order.